AMS SHORT COURSE LECTURE NOTES

Introductory Survey Lectures

published as a subseries of

Proceedings of Symposia in Applied Mathematics

Proceedings of Symposia in
APPLIED MATHEMATICS

Volume 60

Symbolic Dynamics and its Applications

American Mathematical Society
Short Course
January 4–5, 2002
San Diego, California

Susan G. Williams
Editor

American Mathematical Society
Providence, Rhode Island

LECTURE NOTES PREPARED FOR THE
AMERICAN MATHEMATICAL SOCIETY SHORT COURSE
SYMBOLIC DYNAMICS AND ITS APPLICATIONS
HELD IN SAN DIEGO, CALIFORNIA
JANUARY 4–5, 2002

The AMS Short Course Series is sponsored by the Society's Program Committee for National Meetings. The series is under the direction of the Short Course Subcommittee of the Program Committee for National Meetings.

2000 *Mathematics Subject Classification.* Primary 37B10, 37B50, 37A15, 37F45, 94B05, 19C99.

Library of Congress Cataloging-in-Publication Data

American Mathematical Society Short Course on Symbolic Dynamics and its Applications : (2002 : San Diego, Calif.)

Symbolic dynamics and its applications : American Mathematical Society, Short Course, January 4–5, 2002, San Diego, California / Susan G. Williams, editor.

p. cm. — (Proceedings of symposia in applied mathematics, ISSN 0160-7634 ; v. 60)

Includes bibliographical references and index.

ISBN 0-8218-3157-7 (alk. paper)

1. Symbolic dynamics—Congresses. I. Williams, Susan G., 1953– II. Title. III. Series.

QA614.85.A44 2002
514′.74—dc22 2003062891

∞ The paper used in this book is acid-free and falls within the guidelines
established to ensure permanence and durability.
Visit the AMS home page at `http://www.ams.org/`

10 9 8 7 6 5 4 3 2 1 09 08 07 06 05 04

Contents

Preface

The foundation of the field of symbolic dynamics is generally credited to Jacques Hadamard, who used infinite symbol sequences in his analysis of geodesic flow on negatively curved surfaces in 1898. Hadamard's symbolic techniques were soon adopted and extended by other authors. However, the field had to wait forty years for its christening by Marston Morse and Gustav Hedlund, who provided the first systematic study of symbolic dynamical systems as objects of interest in their own right. This rather prescient paper at the dawn of the computer age set the stage for the mathematical analysis of codes and finite-alphabet communication systems using the techniques of dynamics and ergodic theory, most notably in the pioneering work of C.E. Shannon on the mathematical theory of communication. Fifty years after Hadamard applied symbolic techniques to dynamics, Shannon and others were applying dynamical techniques to symbols.

In the fifty-odd years since then, symbolic dynamics has expanded its reach to apply, and be applied to, many areas. It has broken the confines of one dimension to encompass multi-dimensional arrays. The six chapters of this volume provide an introduction to the field as it is studied today and a sampler of its concerns and applications. They are expanded versions of the lectures given in the American Mathematical Society Short Course on Symbolic Dynamics and its Applications held in San Diego on January 4-5, 2002. I would like to take this opportunity to thank Jim Maxwell, Wayne Drady and the other AMS staff members who coordinated the short course and worked behind the scenes to make it run smoothly.

Susan G. Williams

Proceedings of Symposia in Applied Mathematics
Volume **60**, 2004

Introduction to Symbolic Dynamics

Susan G. Williams

ABSTRACT. We give an overview of the field of symbolic dynamics: its history, applications and basic definitions and examples.

1. Origins

The field of symbolic dynamics evolved as a tool for analyzing general dynamical systems by discretizing space. Imagine a point following some trajectory in a space. Partition the space into finitely many pieces, each labeled by a different symbol. We obtain a *symbolic trajectory* by writing down the sequence of symbols corresponding to the successive partition elements visited by the point in its orbit. We may ask: Does the symbolic trajectory completely determine the orbit? Can we find a simple description of the set of all possible symbolic trajectories? And, most important, can we learn anything about the dynamics of the system by scrutinizing its symbolic trajectories? The answers to these questions will depend not only on the nature of our dynamical system, but on the judicious choice of a partition.

Hadamard is generally credited with the first successful use of symbolic dynamics techniques in his analysis of geodesic flows on surfaces of negative curvature in 1898 [**Ha**]. Forty years later the subject received its first systematic study, and its name, in the foundational paper of Marston Morse and Gustav Hedlund [**MH**]. Here for the first time symbolic systems are treated in the abstract, as objects in their own right. This abstract study was motivated both by the intrinsic mathematical interest of symbolic systems and the need to better understand them in order to apply symbolic techniques to continuous systems. However, a further impetus was given by the emergence of information theory and the mathematical theory of communication pioneered by C.E. Shannon [**Sh**].

Symbolic dynamics has continued to find application to an ever-widening array of continuous systems: hyperbolic diffeomorphisms, maps of the interval, billiards, complex dynamics and more. At the same time it contributes to, and finds inspiration in, problems arising in the storage and transmission of data, as we will see in Brian Marcus's chapter. Computer simulations of continuous systems necessarily involve a discretization of space, and results of symbolic dynamics help us

1991 *Mathematics Subject Classification.* Primary 37B10.
Key words and phrases. Symbolic Dynamics.
The author was supported in part by NSF Grant #0071004.

understand how well, or how badly, the simulation may mimic the original. And symbolic dynamics *per se* has proved a bottomless source of beautiful mathematics and intriguing questions.

There are two excellent texts on symbolic dynamics. *An Introduction to Symbolic Dynamics and Coding,* by Douglas Lind and Brian Marcus [**LM**], has the more modest prerequisites (for example, no prior knowledge of topology or measure theory is assumed), while B. Kitchens's more compact *Symbolic Dynamics: One-sided, Two-sided and Countable State Markov Shifts* [**Ki**] assumes basic first-year graduate mathematics. For the most part, we have followed the notation and terminology of [**LM**] in this survey. Also highly recommended is the collection [**BMN**] of survey articles from the 1997 Summer School on Symbolic Dynamics and its Applications in Frontera, Chile. The selection of topics is largely complementary to that of this short course.

2. Two simple examples

Consider the unit interval $I = [0, 1)$ and the map f that sends $x \in I$ to $\{2x\}$, the fractional part of $2x$. We are interested in the orbit x, $f(x)$, $f^2(x) = f(f(x)), \ldots$. If we wanted to trace this orbit on a computer screen we might begin by resolving the interval into 2^{10} pixels. However, we will content ourselves with a much cruder discretization of space: we will break I into just two parts, $I_0 = [0, \frac{1}{2})$ and $I_1 = [\frac{1}{2}, 1)$. We assign to x a symbolic trajectory $x_0x_1x_2\ldots$ where x_i is 0 or 1 according as $f^i(x)$ is in I_0 or I_1. A little consideration will show that the expression $.x_0x_1x_2\ldots$ is simply a binary expansion of the number x. Hence x is completely determined by its symbolic trajectory. We see here an exchange of spatial information for time series information mediated by dynamics: We can recover the complexity of the continuum I from our crude 2-element partition, provided that we observe the evolution of the system for all time.

What symbolic trajectories will appear in this scheme? All binary sequences except those that end in $111\ldots$. This awkward exception can be removed by working instead with closed intervals $I = [0, 1]$, $I_0 = [0, \frac{1}{2}]$ and $I_1 = [\frac{1}{2}, 1]$, and mapping sequences to points instead of the other way around. Beginning with a binary sequence $x_0x_1x_2\ldots$, we can assign to it the unique point

$$x = \bigcap_{i=1}^{\infty} f^{-i}(I_{x_i})$$

that has that symbolic itinerary. Then, for example, $\frac{1}{2}$ will arise from two symbolic trajectories, corresponding to the two binary expansions $\frac{1}{2} = .1000\cdots = .0111\ldots$. It is common practice in the application of symbolic techniques to sacrifice strict one-to-one correspondence for a simpler description of the set of symbolic trajectories.

Our map f now has a very pleasant symbolic representation. If $x = .x_0x_1x_2\ldots$ then $f(x) = \{2x\} = .x_1x_2x_3\ldots$. We shift the symbolic sequence to the left and lop off the initial symbol. The key to the utility of symbolic dynamics is that the dynamics is given by a simple coordinate shift. Dynamic properties that might have seemed elusive in the original setting now become transparent. For instance, we can immediately identify the points of period 3 (that is, with $f^3(x) = x$). They are the eight points with repeating symbolic trajectories $x_0x_1x_2x_0x_1x_2\ldots$. Note that the set of points with symbolic representation beginning with some fixed initial string $x_0x_1\dots x_n$ is the dyadic interval $[k/2^{n+1}, (k+1)/2^{n+1}]$, where $k = x_0\cdot 2^n + x_1\cdot 2^{n-1} + \cdots + x_n$. The orbit of a point is dense in I, visiting every interval no matter how small, if and only if its symbol sequence contains all possible finite strings of 0's and 1's.

As a variation on the first example, consider the map $g(x) = \{\gamma x\}$ on I, where $\gamma = (1+\sqrt{5})/2$ is the golden mean. We let $I_0 = [0, \frac{1}{\gamma}]$ and $I_1 = [\frac{1}{\gamma}, 1]$. Since $\gamma = 1 + \frac{1}{\gamma}$, we have $g(I_0) = I$ and $g(I_1) = I_0$. A point that lands in I_1under some iterate of g must move to I_0 at the next iteration. In fact, it is not hard to see that the set of symbolic trajectories is exactly the set of binary sequences that do not contain the string 11.

The symbolic trajectory $x_0x_1x_2\dots$ corresponds to a series expansion $x = x_0\gamma^{-1} + x_1\gamma^{-2} + x_2\gamma^{-3} + \cdots$. Expansions of numbers with respect to a non-integer base β are called *beta expansions*. There is a very interesting literature relating dynamic properties of symbolic systems obtained by beta expansions to the number-theoretic properties of beta. The chapter by C. Frougny in [**BMN**] provides an up-to-date survey.

3. Full shifts and subshifts

We let $\mathcal{A}$ denote a symbol set or *alphabet*, which for now we assume to be finite. The (two-sided) *full $\mathcal{A}$-shift* is the dynamical system consisting of the set of biinfinite symbol sequences, together with the *shift map* σ that shifts all coordinates to the left. More formally, our space is

$$\mathcal{A}^{\mathbb{Z}} = \{x = (x_i)_{i\in\mathbb{Z}} :\ x_i \in \mathcal{A} \text{ for all } i \in \mathbb{Z}\}$$

and the map $\sigma : \mathcal{A}^{\mathbb{Z}} \to \mathcal{A}^{\mathbb{Z}}$ satisfies $(\sigma x)_i = x_{i+1}$. If $\mathcal{A} = \{0, 1, \dots, n-1\}$ we call $\mathcal{A}^{\mathbb{Z}}$ the full n-shift.

The advantage of working with biinfinite sequences is that the shift map is invertible. However, we may also consider the one-sided $\mathcal{A}$-shift $\mathcal{A}^{\mathbb{N}}$, with the truncating shift map described in the previous section. These arise naturally as symbolic representations of noninvertible maps like the map $x \to \{2x\}$ on I. For simplicity we state most of our definitions for two-sided shifts; the one-sided analogue is generally clear.

We often think of an element of $\mathcal{A}^{\mathbb{Z}}$ as a time series, with x_0 representing the present location or state of our trajectory, $(x_i)_{i<0}$ its past history and$(x_i)_{i>0}$ its future. The action of the shift map is like a tick of the clock, moving us one step into the future.

We consider two points of $\mathcal{A}^{\mathbb{Z}}$ to be close to one another if they agree on a large central block $x_{-n}\dots x_n$ of coordinates. To be more concrete, we can define the distance between distinct points x and y to be $d(x,y) = 2^{-n}$ where n is the smallest integer with $x_{-n} \neq y_{-n}$ or $x_n \neq y_n$. This is a metric, and induces the product topology on $\mathcal{A}^{\mathbb{Z}}$. The map σ and its inverse are continuous: if x and y agree on their central $2n+1$ coordinates, then σx and σy agree at least on their central $2n-1$ coordinates.

A *subshift* or *shift space* is a closed subset of some full shift $\mathcal{A}^{\mathbb{Z}}$ that is invariant under the action of σ. For example, the set of binary sequences that do not contain the string 11 is a subshift of the 2-shift. It is closed because its complement is open: if a sequence contains 11 then every sequence sufficiently close to it does as well. This subshift is often called the *golden mean shift*, in part because of its connection to the golden mean beta expansion. More generally, let $\mathcal{F}$ be any set of finite strings (also called *words* or *blocks*) of symbols of $\mathcal{A}$. The set of sequences that do not contain any word of $\mathcal{F}$ is a subshift $X_{\mathcal{F}}$ of $\mathcal{A}^{\mathbb{Z}}$. In fact, every subshift is of this type, as an easy topological argument will show.

If $X_{\mathcal{F}}$ is determined by a finite set $\mathcal{F}$ of "forbidden" words, we call $X_{\mathcal{F}}$ a *(sub)shift of finite type*, or SFT for short. This is the most fully studied class of symbolic dynamical systems, and the one that has been exploited most in the analysis of general dynamical systems. The systems originally considered by Hadamard were of this type.

For any subshift X we will denote the set of words of length n that appear in some element of X by $\mathcal{B}_n(X)$, the *allowed n-blocks* of X. If $\mathcal{F}$ is a set of words of length not exceeding m, then the SFT $X_{\mathcal{F}}$ is characterized by its set of allowed m-blocks: $\mathcal{B}_m(X)$ is the set of m-blocks over the alphabet $\mathcal{A}$ that do not contain a word of $\mathcal{F}$, and a sequence $x \in \mathcal{A}^{\mathbb{Z}}$ is in $X_{\mathcal{F}}$ if and only if all of its m-blocks are in $\mathcal{B}_m(X)$. A shift of finite type that is determined by its m-blocks is an $(m-1)$*-step* SFT. The idea behind this terminology is that we must look back $m-1$ steps in our symbolic sequence to see which symbols we are allowed to write next. The golden mean shift is a 1-step SFT, with allowed 2-blocks 00, 01 and 10. If we allow consecutive 1's, but no strings of three in a row, we get a 2-step SFT.

A simple example of a subshift that is not of finite type is the *even shift* first studied by B. Weiss [**We**]. It consists of all binary strings in which two 1's are always separated by an even number of 0's. Its set of forbidden words is $\mathcal{F} = \{101, 10001, 1000001, \dots\}$. A variation on this theme is the *prime gap shift*, in which two 1's are separated by a prime number of 0's.

A point of notation is in order before we close this section. We have been speaking of "the"shift map σ on an arbitrary subshift X. In careful parlance, two maps are not the same if they have different domains. A shift space is really a pair (X, σ_X), where X is a closed subset of some $\mathcal{A}^{\mathbb{Z}}$ invariant under the coordinate shift on that particular full shift, and σ_X is the restriction of that coordinate shift to X. In these notes we use X as a shorthand for the pair (X, σ_X), but often the map σ_X is singled out instead.

4. Coding and isomorphism

The term *code* is variously used in symbolic dynamics and related fields for maps of different sorts between symbolic systems, or from a general dynamical system to a symbolic one. For example, the one-sided golden mean shift might be described as a coding of the map $x \to \{\gamma x\}$ on the interval. It should be noted that in coding and information theory, a mapping may be called an *encoder*, and its image a *code*.

Within symbolic dynamics we are naturally interested in maps that preserve, at least to some extent, the topology and dynamics of the shift space. We want nearby points to go to nearby points, and if x is sent to y then its shift σx should go to σy. A *homomorphism* from one subshift to another is a continuous map ϕ that commutes with the shift, that is, for which $\phi \circ \sigma = \sigma \circ \phi$. An onto homomorphism is traditionally called a *factor map*, a term also used in ergodic theory, although the term *quotient map* would be more consistent with usage in other areas of mathematics. An isomorphism (invertible homomorphism) from one subshift to another is also called a *topological conjugacy*, or simply a *conjugacy*, as is usual in the general theory of dynamical systems.

We can define a factor map ϕ from the golden mean shift to the even shift as follows: map $x = (x_i)$ to $y = (y_i)$ where $y_i = 1 - (x_i + x_{i+1})$ for all i. Since each 1 in x is immediately preceded and followed by a 0, the 0's in y are produced in pairs. Clearly $\sigma\phi(x) = \phi(\sigma x)$. The map is continuous because it is given by a local rule, so that a central block of y is determined by a slightly longer central block of x. In general, a *sliding block code* from a subshift X to a subshift Y is a map ϕ given by a local rule $(\phi(x))_i = \Phi(x_{i-m} \dots x_{i+a})$, where m and a are integers with $-m \leq a$ and Φ is a map from the $(m + a + 1)$-blocks of X to the symbols of Y. The numbers m and a, usually taken to be nonnegative, are called respectively the *memory* and *anticipation* of the code. An argument using the compactness of X yields the Curtis-Hedlund-Lyndon theorem: every homomorphism between subshifts is given by a sliding block code.

Of particular interest are the *higher block codes*. We can define a homomorphism from any subshift X into the full $\mathcal{B}_n(X)$-shift by the sliding block code $\Phi(x_0x_1 \dots x_{n-1}) = [x_0x_1 \dots x_{n-1}]$. Here we use the square brackets to emphasize that the enclosed block is being treated as a single symbol in a new alphabet. Thus when $m = 2$, the sequence $\dots x_{-1}x_0x_1x_2 \dots$ is sent to

$$\dots [x_{-1}x_0][x_0x_1][x_1x_2] \dots .$$

This homomorphism is clearly one-to-one. Its image is the *n-block presentation* of X, denoted $X^{[n]}$.

Higher block presentations provide an important technical tool in symbolic dynamics. If X is an m-step shift of finite type, then $X^{[m]}$ is a 1-step SFT: the allowed 2-blocks of $X^{[m]}$ are the blocks $[x_0 \dots x_{m-1}][x_1 \dots x_m]$ where $x_0 \dots x_m$ is an allowed $(m + 1)$-block of X. Thus every SFT is conjugate to a 1-step SFT. Also, if ϕ is a sliding block code with memory m and anticipation l as described above, it induces a sliding block code ψ of zero memory and anticipation from $X^{[m+l]}$ to Y given by the block map $\Psi([x_{i-m} \dots x_{i+l}]) = \Phi(x_{i-m} \dots x_{i+l})$ from symbols (1-blocks) of $X^{[m+l]}$ to symbols of Y. By this device we are often able to reduce general arguments about sliding block codes to the case of one-block codes $(\phi(x))_i = \Phi(x_i)$.

5. Graphs and matrices

Recall that a 1-step SFT X is characterized by its set of allowed 2-blocks, that is, by a list of which symbols may follow which in our symbol sequences. We may represent such a system by a directed graph: the vertices are the symbols of the alphabet $\mathcal{A}$ and there is an edge from a to b if and only if the word ab is allowed. Each element (x_i) of X corresponds to a biinfinite walk on the graph, following edges from one vertex to the next. Conversely, a (finite) directed graph G with no parallel edges determines a 1-step SFT $\hat{X}_G$ with alphabet equal to the set of vertices of G. We call this the *vertex shift* associated with G. The graph of the golden mean shift is shown in figure 1. From here on, *graph* will always mean a directed graph.

A graph G with n vertices is conveniently described by giving its *adjacency matrix*, the $n \times n$ matrix $A = (a_{ij})$ where a_{ij} is the number of edges from the ith vertex to the jth. Thus every vertex shift corresponds to a square matrix A of 0's

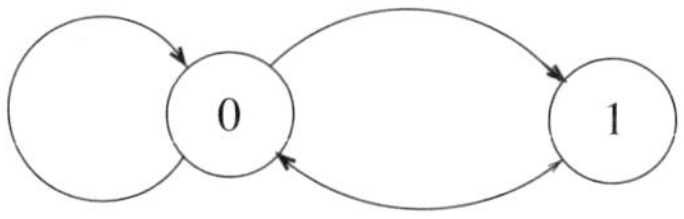

FIGURE 1. Vertex graph of the golden mean shift

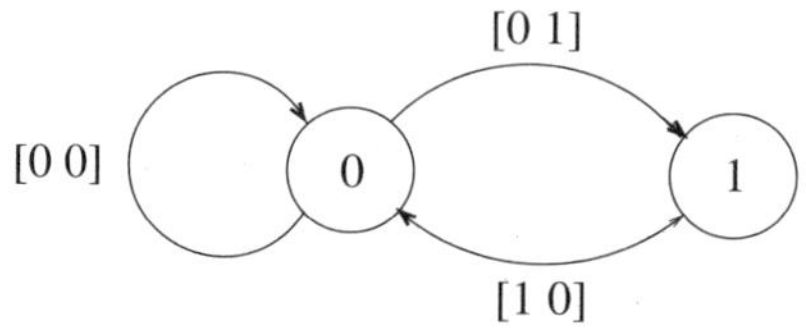

FIGURE 2. 2-block presentation of the golden mean shift

and 1's, sometimes called the *transition matrix* of the vertex shift. The transition matrix for the golden mean shift is

$$\begin{pmatrix} 1 & 1 \\ 1 & 0 \end{pmatrix}.$$

Vertex shifts capture the constraints of a 1-step SFT in an appealingly simple way. Graphs of this sort are used in the field of stochastic processes to describe Markov chain models. The vertices are states of a system, and an edge from a to b is labeled by the probability of transition from state a to state b, which is assumed to be stationary and independent of previous states. The absence of an edge from a to b indicates zero transition probability. Then the vertex shift given by the graph is the underlying topological space supporting the Markov chain. For this reason 1-step shifts of finite type are also called *topological Markov chains.*

Even if the graph G has parallel edges, we can still view it as representing a 1-step SFT if we take the set of edges of G, instead of the vertices, as the symbol set. The *edge shift* X_G is the set of biinfinite walks on the edges of G, that is, the set of edge sequences (x_i) such that the terminal vertex of x_i is the initial vertex of x_{i+1} for all i. The edge shift is also denoted byX_A, where A is the adjacency matrix of G as before.

Edge shifts allow a more efficient representation in many cases. For example, the full n-shift is given as a vertex shift by the $n \times n$ matrix of 1's, and as an edge shift by the 1×1 matrix (n). (The full shift is a shift of finite type, with empty set of forbidden words.) Not every 1-step SFT is an edge shift. For example, there is no graph with two edges that represents the golden mean shift. However, if X is the vertex shift with graph G then its 2-block presentation $X^{[2]}$ can be naturally identified with the edge shift of G by identifying an edge from vertex a to vertex b with the 2-block $[ab]$. Hence every shift of finite type is conjugate to an edge shift. In figure 2 we show the 2-block presentation of the golden mean shift represented as an edge shift.

It can be seen that a subshift that is conjugate to a shift of finite type is itself of finite type. However, the homomorphic image of an SFT may not be an SFT, as we can see by the factor map from the golden mean shift to the even shift. The class of subshifts that are factors of SFT are called the *sofic* shifts. This term, derived from the Hebrew word for *finite*, was coined by B. Weiss [**We**]. Suppose

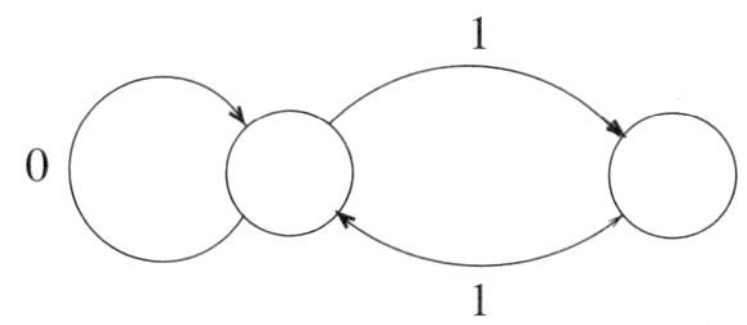

FIGURE 3. Graph of even shift

Y is a factor of an SFT X. By replacing X with a higher block presentation if necessary, we can assume that X is an edge shift X_G and that the factor map is a 1-block code. That is, it takes an edge sequence $x = (x_i) \in X$ to $y = (\Phi(x_i)) \in Y$ where Φ is a map from edges of G to symbols of Y. If we label each edge e of G with the symbol $\Phi(e)$, then the image of an edge sequence under the sliding block code is the sequence of edge labels. Hence any sofic shift may be represented by a labeling of the edges of a directed graph with symbols (not necessarily distinct) from some alphabet $\mathcal{A}$. As with SFT, the elements of Y correspond to biinfinite walks in this graph.

The 2-block code from the golden mean shift to the even shift described in the previous section produces the edge labeling in figure 3. It is easy to see that the prime gap system cannot be represented by a finite labeled graph, so it is not a sofic shift.

Sofic systems have a natural connection to automata theory. We think of the vertices of the graph as internal states of a machine, and the label on an edge from v to v' as the instruction that the machine, when in state v, should go to state v' if that label is read as input. A *nondeterministic finite-state automaton* (NFA) is just such a directed, edge-labeled graph, with one or more designated *initial states* and *accepting states*. The automaton is said to *accept* or *recognize* a word $b_1 \dots b_n$ if this word labels a path from an initial state to an accepting state. The set of all words accepted by an NFA is a *regular language*. In this terminology, the set of allowed blocks of all lengths in a sofic system Y is a regular language given by a NFA in which all the states are both initial and accepting. For more on connections between automata theory and symbolic dynamics, see [**BP**] or [**BMN**].

6. Invariants

One of the most basic questions we may ask about symbolic dynamical systems is how we can tell when two subshifts are conjugate. Even for the simplest class of systems, the shifts of finite type, a complete and effective classification remains elusive.

On the other hand, we have many ways of telling that two subshifts are different. By an *invariant* of conjugacy we mean any quantity or mathematical object that we can assign to subshift that remains unchanged when we replace the subshift by a conjugate one. There are several well-known invariants that can be defined for dynamical systems in general, and others that apply only to shift spaces, or to the smaller class of SFT. We list a few:

6.1. Periodic point count. If f is a homeomorphism of a compact space X, we denote by $\mathrm{Fix}_n(X)$ the set of all $x \in X$ with $f^n(x) = x$. We will call these the period n points of the dynamical system. Note that with this terminology, a period

n point is also a period m point if m is a multiple of n. We distinguish the smallest such n by calling it the least period of x. A conjugacy between dynamical systems preserves periods of points, so conjugate shifts have the same cardinality of period n points for every n.

As we observed before, for a shift space the period n points are those sequences (x_i) with $x_{i+n} = x_i$ for all i. If the alphabet $\mathcal{A}$ has r symbols then there are at most r^n period n points.

It is a well-known fact that in a graph with adjacency matrix A, the number of walks of length n from vertex i to vertex j along the edges of the graph is just the ij entry of A^n. For an edge shift X_A the period n points correspond to walks of length n with the same initial state and final state i. Hence the number of period n points is easily computed as the trace, or sum of the diagonal entries, of A^n.

If the number $p_n(X)$ of period n points of X is finite for all n, the periodic point count can be encoded in the Artin-Mazur *zeta function* of X. This is the power series

$$\zeta_X(t) = \exp\Big(\sum_{n=1}^{\infty} \frac{p_n(X)}{n} t^n\Big).$$

For the SFT X_A given by an $r \times r$ transition matrix A, this function takes the remarkably simple form

$$\begin{aligned} \zeta_A(t) &= \frac{1}{\det(I - tA)} \\ &= \frac{1}{t^r \chi_A(t^{-1})}, \end{aligned} \tag{6.1}$$

where χ_A is the characteristic polynomial of A. This formula, known as the Bowen-Lanford formula, can be verified by putting A in its Jordan normal form and using the trace result cited above. More generally, it follows from a theorem of Manning [**Ma**] that every sofic system has rational zeta function.

6.2. Topological entropy. Topological entropy was defined in [**AKM**] for general compact dynamical systems, in analogy with the concept of measure theoretic entropy developed earlier by Shannon and by Kolmogorov and Sinai. We will not give the general definition, which is fairly involved, since there is a much simpler formulation for the special case of symbolic dynamical systems. This formulation can be motivated in terms of information theory.

Think of an allowed n-block of a subshift X as the information we would gain by observing our symbolic dynamical system for n ticks of the clock. If X is the full 2-shift, the n-block may be any one of the 2^n binary strings of length n, so by recording what particular n-block occurs we gain n bits of information, or one bit per symbol. However, if X is the golden mean shift then not all binary strings can occur. We gain less information from observing a particular block since there are many blocks that we could have ruled out in advance. The number N_n of binary strings of length n with no consecutive 1's satisfies the Fibonacci recurrence relation

$$N_{n+2} = N_{n+1} + N_n,$$

since we can form an allowed $(n + 2)$-block either by tacking a 0 onto an allowed $(n + 1)$-block or by putting 01 after an annowed n-block. The number N_n is the $(n + 2)$-th Fibonacci number, and grows asymptotically as $C\gamma^n$ where C is a constant and $\gamma = (1 + \sqrt{5})/2$ is the golden mean. We may say that the amount of information we gain by observing a particular n-block of the golden mean shift is about $\log_2(C\gamma^n) = n\log_2\gamma + \log_2 C$ bits, or roughly $\log_2(\gamma)$ bits per symbol.

We define the (topological) *entropy* of a shift space X to be the limit

$$h(X) = \lim_{n\to\infty}\frac{1}{n}\log N_n$$

where $N_n = N_n(X)$ is the number of allowed n-blocks of X. We can describe $h(X)$ as the per-symbol information rate of the shift, or as the exponential growth rate of the number of n-blocks. That this limit exists (and is equal to the infinum of the sequence) can be established from the observation that $N_{m+n} \le N_m \cdot N_n$. Whether we use natural or base 2 logarithms is a matter of personal proclivity; with the natural log we are measuring information in *nats* instead of bits.

If Y is a factor of X given as the image of a sliding m-block code, then the number of n-blocks of Y cannot exceed the number of $(n+m)$-blocks of X. Thus

$$\begin{aligned} h(Y) &= \lim_{n\to\infty}\frac{1}{n}\log N_n(Y) \\ &\le \lim_{n\to\infty}\frac{1}{n}\log N_{n+m}(X) \\ &= \big(\lim_{n\to\infty}\frac{n+m}{n}\big)h(X) = h(X). \end{aligned} \tag{6.2}$$

Since conjugate shifts are factors of one another, they have equal entropy.

We see immediately that the entropy of the full r-shift is $\log r$, since there are r^n words of length n. For a shift of finite type X_A given by a transition matrix A, the number of allowed n-blocks is the sum of the entries of A^n. By a result known as the Perron-Frobenius theorem, every square nonnegative matrix A has a nonnegative real eigenvalue λ_A (the *Perron-Frobenius eigenvalue*) that is greater than or equal to the modulus of every other eigenvalue of A. For the golden mean shift $\lambda_A = \gamma$. It can be shown that in general, $h(X_A) = \log\lambda_A$.

6.3. Algebraic invariants for shifts of finite type. As we have seen, for a shift of finite type X_A the number of period n points, the zeta function and the entropy can all be simply expressed in terms of algebraic invariants of the transition matrix A. We can see from the invariance of the zeta function that if X_A and X_B are conjugate SFT then A and B must have the same characteristic polynomial, up to some factor t^k. In fact a stronger statement is true: the Jordan forms of the invertible parts of A and B must be the same.

Another useful algebraic invariant is the *Bowen-Franks group*. If A is an $r\times r$ transition matrix, its Bowen-Franks group is the quotient of $\mathbb{Z}^r$ by its image under the matrix $I - A$:

$$BF(A) = \mathbb{Z}^r/\mathbb{Z}^r(I-A).$$

If X_A is conjugate to X_B then the Bowen-Franks groups of A and B must be isomorphic [**BF**]. This condition is easily checked by computing the elementary divisors of A and B. The Bowen-Franks group is, in fact, invariant under flow equivalence of SFT, a weaker equivalence than conjugacy.

One of the most sought-after goals in symbolic dynamics has been a complete and effective classification of shifts of finite type in terms of their transition matrices. In 1973 R. Williams [**Wi**] introduced two important equivalence relations on the set of square matrices over the non-negative integers, *shift equivalence* and *strong shift equivalence*. He showed that finite type shifts X_A and X_B are conjugate if and only A and B are strong shift equivalent. Strong shift equivalence implies shift equivalence; the converse statement became known as the shift equivalence conjecture or Williams conjecture. The importance of the conjecture lies in the fact that latter equivalence is more tractable: in fact, it is known to be decidable [**KR1**]. Jack Wagoner's chapter relates the developments that led in 1997 to a counterexample to the shift equivalence conjecture by K. H. Kim and F. Roush [**KR2**] following joint work with Wagoner, and outlines the current state of affairs.

Crucial to the solution of the shift equivalence problem was the study of the group of automorphisms, or self-conjugacies, of a shift of finite type. Bob Devaney's chapter will take us a step beyond Hadamard's inspiration by showing how the automorphism group of a shift can encode information from other dynamical settings, in this case families of complex polynomial maps.

7. Wider vistas

There are several ways in which we can relax our notion of symbolic dynamical system to get a larger class of systems. One is to allow a countable alphabet in place of a finite one. This makes the shift space noncompact, which introduces some complications, for example in finding an appropriate definition of entropy. The theory of countable state topological Markov chains—vertex shifts on a graph with countably many vertices but only finitely many edges entering or leaving each edge—is of particular interest. A good introduction to this topic is Chapter 7 of [**Ki**].

We could instead choose our symbol set to be a compact group such as the circle $\mathbb{T} = \mathbb{R}/\mathbb{Z}$. Although this takes us far from the original idea of a symbolic dynamical system as a discretization of space, if our map is still a coordinate shift map then some of the spirit and techniques remain. We will encounter shift spaces with alphabet $\mathbb{T}$ in Doug Lind's chapter.

The study of a dynamical system (X, σ) is really the study of the behavior of X under the iterates σ^n of σ. We may describe this as an action of the group $\mathbb{Z}$ on X: for every $n \in \mathbb{Z}$ we have a coordinate shift map σ^n that shifts all coordinates by n, with $\sigma^{m+n} = \sigma^m \circ \sigma^n$. In general, an *action* of a group G by homeomorphisms on a space X is a map that takes each $g \in G$ to a homeomorphism f_g of X in such a way that $f_{gh} = f_g \circ f_h$. Another way to broaden our notion of symbolic dynamical systems is to consider actions by other discrete infinite groups in place of $\mathbb{Z}$. One of the most exciting currents in symbolic dynamics is development of the theory of $\mathbb{Z}^d$-actions. Elements of a $\mathbb{Z}^d$ symbolic dynamical system are d-dimensional arrays $(x_{\mathbf{n}})_{\mathbf{n}\in\mathbb{Z}^d}$ of symbols, and for each $\mathbf{m} \in \mathbb{Z}^d$ there is a shift map $\sigma^{\mathbf{m}}$ that shifts all coordinates by $\mathbf{m}$. Doug Lind's chapter is a survey of these multidimensional systems. As you will see, there are very nice results for special class of multidimensional systems with algebraic structure, but the study of general $\mathbb{Z}^d$-actions involves substantial complications not found in the one-dimensional case.

We can think of an element of a two-dimensional symbolic dynamical system as a tiling of the plane by unit square tiles of different colors, where the colors are simply our alphabet of symbols. A translation of the tiling by an integer vector $\mathbf{m}$ gives another tiling that represents a coordinate shift of the original. To bring the techniques of symbolic dynamics to bear on the general problem of tiling the plane with tiles of various shapes we need to allow general translations in the plane. In Robbie Robinson's chapter, which examines not just planar but d-dimensional tilings, we will be working in the setting of $\mathbb{R}^d$ actions by translation of Euclidean space.

References

[AKM] R. Adler, A. Konheim & M. McAndrew, *Topological Entropy* **114** (1965), 309–319.

[BF] R. Bowen & J. Franks, *Homology for zero-dimensional basic sets*, Annals of Math. **106** (1977), 37–92.

[BP] M.-P. Béal & D. Perrin, *Symbolic dynamics and finite automata*, Handbook of Formal Languages (G. Rozenberg & A. Salomaa, eds.), vol. 2, Springer-Verlag, Berlin, 1997, pp. 463–505

[BMN] F. Blanchard, A. Maass & A. Nogueira, eds., *Topics in Symbolic Dynamics and Applications*, Cambridge Univ. Press, Cambridge, U.K., 2000.

[Bo] M. Boyle, *Symbolic dynamics and matrices*, Combinatorial and Graph Theoretic Problems in Linear Algebra, IMA Volumes in Math. and its Appl. (R. Brualdi et al., eds.), vol. 50, 1993, pp. 1–38.

[Ha] J. Hadamard, *Les surfaces à courbures opposées et leur lignes géodésiques*, J. Maths. Pures et Appliqués **4** (1898), 27–73.

[Ki] B. Kitchens, *Symbolic Dynamics: One-sided, Two-sided and Countable State Markov Chains*, Springer-Verlag, Berlin, Heidelberg, 1998.

[KR1] K.H. Kim & F.W. Roush, *Decidability of shift equivalence*, in J.W. Alexander, ed., Dynamical systems, Lecture Notes in Mathematics, vol. 1342, Springer-Verlag, Heidelberg, 1988.

[KR2] K.H. Kim & F.W. Roush, *Williams's conjecture is false for irreducible subshifts*, Elec. Res. Announc. Amer. Math. Soc. **3** (1997), 105-109.

[MH] M. Morse & G.A. Hedlund, *Symbolic Dynamics*, Amer. J. Math. **60** (1938), 815–866.

[LM] D. Lind & B. Marcus, *Symbolic Dynamics and Coding*, Cambridge Univ. Press, Cambridge, U.K., 1995.

[Ma] A. Manning, *Axiom A diffeomorphisms have rational zeta functions*, Bull. London Math. Soc. **3** (1971), 215–220.

[Sh] C.E. Shannon, *A mathematical theory of communication*, Bell Syst. Tech. J. **27** (1948), 379–423, 623–656.

[We] B. Weiss, *Subshifts of finite type and sofic systems*, Monats. Math. **77** (1973), 462–474.

[Wi] R.F. Williams, *Classification of subshifts of finite type*, Annals of Math. **98** (1973), 120–153; *erratum*, Annals of Math.**99** (1974), 380–381.

DEPARTMENT OF MATHEMATICS AND STATISTICS, UNIVERSITY OF SOUTH ALABAMA, MOBILE, AL 36688

E-mail address: `swilliam@jaguar1.usouthal.edu`

Proceedings of Symposia in Applied Mathematics
Volume **60**, 2004

Combining Modulation Codes and Error Correcting Codes

Brian Marcus

ABSTRACT. In this talk, we describe modulation codes and error correction codes (ECC), used in data communications and recording systems. We show how ideas in symbolic dynamics have been used to construct and analyze modulation codes as well as to address certain issues that arise when modulation codes and ECC are combined in a system.

1. Introduction

In the abstract, a channel is a "black box" with inputs and outputs. The inputs represent messages that are transmitted through the channel. The outputs are supposed to faithfully represent the inputs. However, distortions in the channel can adversely affect the output. For this reason, coding is applied to protect the messages.

One usually thinks of a channel as a communications system, in which information is sent from one point in space to another. Examples of communications systems include telephones, cellphones, digital subscriber lines and deep space communications. But recording systems, such as magnetic/optical disk/tape drive systems, can also be viewed as channels: data is written at one point in time and read at a subsequent point in time.

Current recording applications require storage devices to have very high immunity against errors. On the other hand, the ever-growing demand for storage forces the designers of such devices to write more data per unit area, thereby making the system less reliable. In magnetic recording systems, this is manifested in the effects of interference between successive transitions in magnetization (called intersymbol interference), inaccurate clocking, and random noise.

A modulation encoder, also known as a constrained encoder or line encoder, transforms an arbitrary user data sequence into a coded sequence that satisfy a given constraint. Typically, the user data sequence is decomposed into short words (e.g., bytes) and encoded into codewords, which, when concatenated together, form the coded sequence. In the most general terms, the purpose of a modulation code is to improve the performance of the system by matching the characteristics of the recorded signals to those of the channel; the recorded signals are thereby constrained in such a way as to reduce the likelihood of error. For instance, run length constraints, which bound the runs of zeros in an encoded data stream, help to mitigate the problems of intersymbol interference and inaccurate clocking (see sections 2.1 and 2.2).

In addition to modulation coding, an error-correction code (ECC) is used to protect the data against random noise sources (some simple examples are given in section 3). The ECC encoder encodes user data words into codewords all of the same length. The hallmarks of a good ECC are (1): any two distinct codewords differ in sufficiently many positions, so as to be distinguishable even after being subjected to a certain amount of channel noise and (2): there are many codewords, so that many different messages can be transmitted. While both error-correction coding and modulation coding have been active for fifty years, ECC enjoys much greater notoriety.

What is the difference between an error-correction code and a modulation code? One difference is that the "goodness" of an error-correction code is measured by how the different codewords relate to one another (e.g., in how many bit locations must any two distinct codewords differ?), whereas the "goodness" of a modulation code is measured by properties of the individual codewords (e.g., how well does each codeword pass through the channel undisturbed?).

On the other hand, this distinction is not hard and fast. Clearly, if an error-correction code is to have any value at all, then its codewords cannot be completely arbitrary and therefore must be constrained. Conversely, in recent years, there has been a great deal of interest in modulation codes that also have error-correction power. Such developments have contributed to a blurring of the lines between these two types of coding. Nevertheless, each subject has its own emphases and fundamental problems that are shaped by the distinction posed in the preceding paragraph.

Ideally the messages transmitted through a channel should be determined by a single code that has both "pairwise" error-correction properties as well as "individual" modulation properties. In this talk, we will survey some methods for combining modulation codes with error correction codes. There is a vast literature on this subject. We will focus on a special case: we will use two different encoders, one for the modulation code and one for the ECC code, and discuss different ways of concatenating the encoders so as to guarantee that the recorded sequences satisfy the modulation constraints and have ECC power. Before proceeding to these issues, we give some brief background on modulation codes in section 2 and on ECC in section 3.

2. Modulation Codes

2.1. Magnetic Recording. In this section, we motivate the idea of a modulation code via an application to magnetic recording. For more background on magnetic recording, see [**WT99**].

The essential components of a magnetic recording system are a recording head, a read-head and a recording medium, such as a rotating magnetic disk. The disk is divided into concentric tracks. To write a string of data along a track, the recording head is positioned above the track, and current sent through the head magnetizes the track in one of two directions called magnetic polarities. This process is illustrated in Figure 1. A clock runs at a constant bit period of T seconds, and at each clock tick (i.e., at each multiple of T), the recording head has the opportunity to change the polarity on the disk: a "1" is recorded by changing the direction of the current, while a "0" is recorded by *not* changing the direction of the current (i.e., doing nothing). In this way, a "1" is represented as a transition in magnetic polarity along a data track, while a "0" is represented as an absence of such a transition. So, we can think of any binary sequence as a sequence of transitions/no-transitions; when we want to emphasize that we are thinking of a binary sequence in this way, we will call it a *transition sequence*.

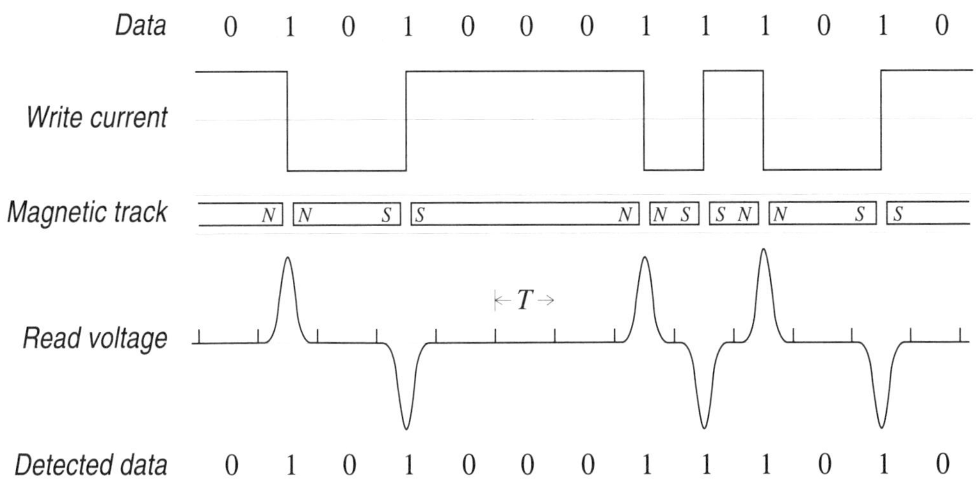

FIGURE 1. Digital magnetic recording (Reprinted from [**LM95**] with permission of Cambridge University Press)

In reading mode, the read-head, positioned above a track on the rotating disk, responds to a magnetic transition by an induced voltage; naturally, the absence of a transition produces no voltage response in the read-head. A bit cell is a window of length T centered on a clock tick. When a sufficiently high (positive or negative) voltage peak is detected within a bit cell, a "1" is declared; otherwise, a "0" is declared. This scheme is known as *peak detection*. Note that the readback voltages corresponding to successive transitions are of opposite signs (see Figure 1), but in most implementations the peak detector ignores the signs.

Now, if successive 1's are too close together, then the voltage responses corresponding to the magnetic transitions may interfere with one another; this is known as *intersymbol interference*. From Figure 2, which shows a reconstruction of the received signal with interference, it is evident that such interference may degrade the signal so that either one or both transitions are missed. In such a case, a recorded "1" will be mis-read as a "0"; perhaps, even worse, the bit cell corresponding to a transition may be incorrectly identified so that a recorded "01" will be mis-read as "10"; such an error is called a *peak shift* or *bit shift*. The likelihood of these kinds of errors can be reduced if 1's are separated sufficiently far apart, or equivalently if any two 1's are separated by a sufficiently long run of 0's.

Determination of the correct runlength of 0's between two successive 1's depends critically on accurate clocking. But accurate clocking can well be compromised by various imperfections in the recording system, such as variations in speed of the rotating disk. Thus, clocking needs to be adjusted periodically via a timing

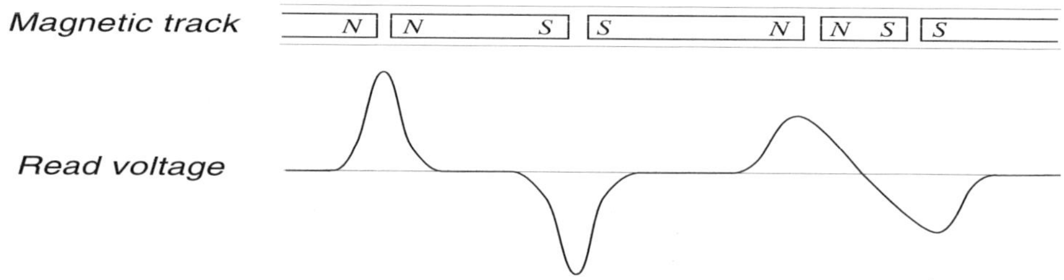

FIGURE 2. Intersymbol interference (Reprinted from [**LM95**] with permission of Cambridge University Press)

control scheme, which adjusts the clock in response to identification of peaks in the received signal. For instance, if a peak is supposed to occur at exactly the mid-point of the bit cell, but instead occurs near the end of the bit cell, then the next clock tick should be delayed. But during a long run of 0's, the ideal received signal will not contain any peaks. Hence, for timing control, it is desirable to avoid long runs of 0's.

2.2. Sofic shifts as channel constraints. The foregoing discussion suggests that it may be helpful to impose constraints on the binary sequences that are actually recorded. Most constraints that arise in practice can be described as sofic shifts: a *sofic shift* is the set of all bi-infinite sequences defined by reading the labels of a labeled finite directed graph (G, L). Here, G is a finite directed graph and L is a labeling of the edges of G. Since sofic shifts often arise as constraints designed to improve the performance of a recording or communications system, the terms *constraint* or *constrained system* or *modulation constraint* are often used, instead of sofic, in the engineering literature. Typically, these alternative terms refer to the corresponding set of finite sequences instead of infinite or bi-infinite sequences.

We will write finite sequences (also called *blocks* or *words*) in the form: $w = w_0 \dots w_{\ell-1}$. A block of length ℓ will be called an ℓ-block or ℓ-word. And $I \xrightarrow{a} J$ signifies an edge in a labeled graph from state I to state J with label a. For more background on sofic shifts, symbolic dynamics and the relationship to modulation coding, see [**Kit98**], [**LM95**], [**MRS98**].

As an example, consider the *runlength-limited (RLL)(d, k) shift*, which is the set of all bi-infinite binary sequences for which 1's occur infinitely often to the left and right and the number of 0's between successive occurrences of a 1 is in between d and k. According to the discussion in Section 2.1, for binary transition sequences the d-constraint reduces the effect of intersymbol interference and the k-constraint aids in timing control.

Many commercial systems, such as magnetic tape recording systems, use the constraint $(d, k) = (1, 7)$ or $(d, k) = (2, 7)$. An example of a portion of a sequence satisfying the $(d, k) = (2, 7)$ shift is

$$\mathbf{w} = 00100001001000000010 .$$

Other recording standards include the $(1, 3)$-RLL shift, which can be found in older floppy disk drives, and the $(2, 10)$-RLL shift, which appears in the compact audio disk and the digital versatile disk (DVD) [**Imm99**].

The RLL(d,k) shift is indeed sofic: it is presented by the labeled graph shown in Figure 3. For each $i = 0,\ldots,k-1$, there is an edge $i \xrightarrow{0} i+1$, and for each $j = d,\ldots,k$, there is an edge $j \xrightarrow{1} 0$.

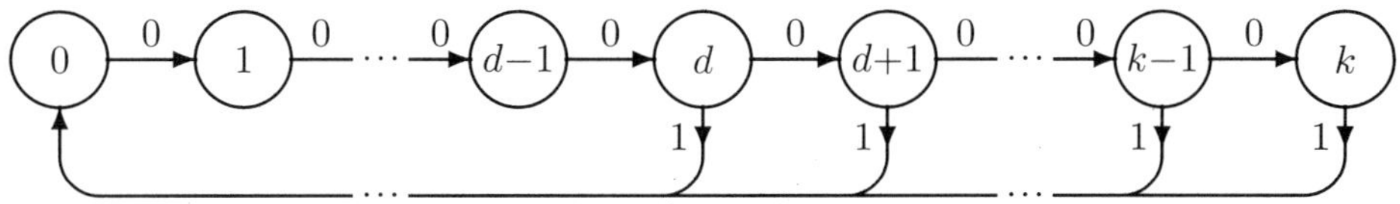

FIGURE 3. Graph presentation of the RLL(d,k) shift.

Note that the RLL(d,k) shift is actually a *shift of finite type*, i.e. a shift space determined by a finite list of forbidden blocks. For instance, the RLL$(1,3)$ shift is the set of sequences that do not contain any element of the list:

$$\{11, 0000\},$$

as a substring of contiguous symbols. However, there are many sofic shifts of interest in practice that are not shifts of finite type. For instance, for an integer $c > 0$, consider the *c-charge-constrained shift*, which is defined as the set of all sequences over the alphabet $\{+1,-1\}$, such that for every block, the algebraic sum of the $+1$'s and -1's lies in between $-c$ and c. Such a shift is indeed sofic, as indicated by the labeled graph of Figure 4. Such constraints are useful in many applications, such as fiber optic links [**WF83**]. They are also used in magnetic recording where the symbols $+1$, -1 refer to magnetization polarity (as opposed to the binary symbols which refer to the presence or absence of a magnetic transition). Moreover, sofic shifts that combine both runlength limits and charge constraints have been used in magnetic recording [**Pat75**] as well as in optical recording [**Imm01**] (in order to define a sofic shift that combines the constraints, one must transform the run length constraint from the $\{0,1\}$ (transition) domain to an equivalent constraint in the $\{-1,+1\}$ (polarity) domain).

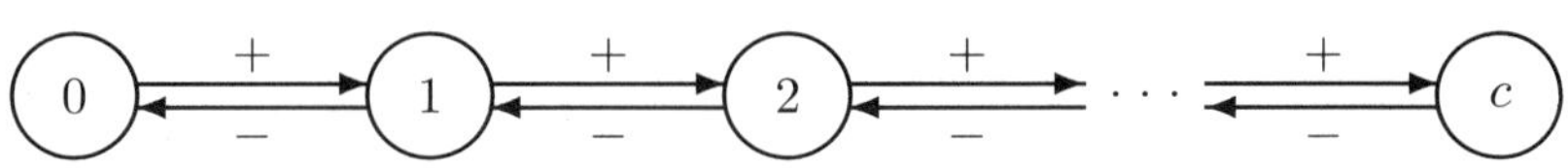

FIGURE 4. c-charge constrained shift

Note that the labeled graphs that we have presented so far are *right resolving*: at each state, all outgoing edges carry distinct labels. It can be shown (using a construction from automata theory) that any sofic shift can be presented by a right resolving graph. Sometimes it will be useful to consider a weaker property, called *right closing*: there is an integer D such that whenever two paths of length $D+1$ have the same initial state and labeling, then they must have the same initial edge. Note that the special case $D = 0$ reduces to right resolving.

There are many constraints of interest in applications beyond those that we have described so far. For instance, Maximum Transition Run (MTR) constraints impose an upper bound on the runlengths of 1's rather than 0's [**MB96**] . In addition, asymmetric RLL constraints, where different restrictions are imposed on the runlengths of 0's and 1's, have been used in optical recording [**UB88**]. Finally, constraints on 2-dimensional patterns, which naturally define 2-dimensional shifts of finite type, have been used in holographic recording [**Mar00**].

2.3. Encoding and Decoding. Once we have a specific sofic shift at hand, we need to encode arbitrary data sequences into sequences that satisfy the constraints of the sofic shift. Such an encoder is often called a *modulation encoder.*

The conceptually simplest modulation encoder is a *rate $p : q$ block encoder* which is a one-to-one correspondence between user p-blocks to coded q-blocks (of course, care must be taken to ensure that the constraints of the sofic shift are satisfied across coded block boundaries). However, greater flexibility can be achieved if one incorporates state dependence in the encoding process. We make this precise as follows.

Let S be a sofic shift and n a positive integer. A *finite-state* (S, n)*-encoder* (sometimes called a *finite-state code*) is a graph $\mathcal{E}$ which has two labelings: an input labeling I over the alphabet $\{0, \ldots, n-1\}$ and an output labeling O over the alphabet of S such that:

(1) each state has exactly n outgoing edges, and they carry the n distinct input labels $\{0, \ldots, n-1\}$ (so, $(\mathcal{E}, I)$ is right resolving) – and –
(2) each sequence generated by the labeled graph $(\mathcal{E}, O)$ belongs to S, and $(\mathcal{E}, O)$ is right closing.

The first condition guarantees that we can encode any n-ary sequence, and the second condition guarantees that the encoded sequences satisfy the constraints of S and that we can decode any sequence, provided that we know its initial state in $\mathcal{E}$.

Suppose that, for some q, S itself is the *q-th power shift* of another sofic shift S', i.e., S is the version of S' obtained by rewriting each element of S' as a concatenation of non-overlapping blocks of length q. Suppose also that $n = 2^p$ for some p. Then an (S, n)-encoder can rightfully be called a *rate $p : q$ finite-state encoder into S'*. An example of such an encoder is given as follows.

EXAMPLE 2.1. Figure 5 depicts a rate $2 : 3$ encoder for the RLL$(0, 1)$ shift. There are two states, A and B. Each state has exactly four outgoing edges. Each edge has two labels, a 2-bit input label and a 3-bit output label. Note that at each state, the set of four outgoing edges carry all possible 2-bit input labels exactly once.

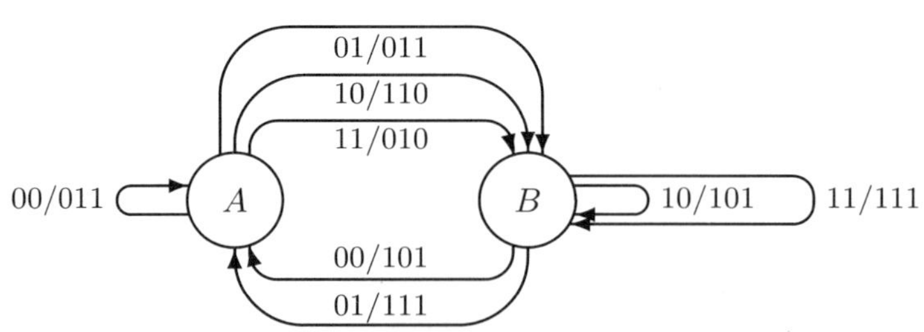

FIGURE 5. Rate $2 : 3$ two-state encoder for RLL$(0, 1)$ sofic shift.

One encodes as follows. First, fix an an initial state, say state A. To encode an arbitrary 2-bit input uv, first find the unique outgoing edge e, with input label uv, outgoing from state A; the encoded 3-codeword is the output label of edge e. To encode the next 2-bit input, execute the same procedure starting from the terminal state of e. For instance, starting at state A, the sequence:

01 01 00 10 10 01 00 00

encodes to:

011 111 011 110 101 111 011 011,

with corresponding state sequence:

$$A\ B\ A\ A\ B\ B\ A\ A\ A.$$

The encoded sequences do indeed satisfy the RLL(0, 1) shift, i.e., you can never see two consecutive 0's; this follows from the observations: (1) you can never see two consecutive 0's on an edge output label, and (2) whenever an edge carries an output label that ends in 0, the following output labels all begin with 1. □

While any finite-state encoder is decodable (by using state information), it is usually required that the decoder have the following additional property: there are integers m and a, such that whenever two paths $e_{-\mathsf{m}}e_{-\mathsf{m}+1}\dots e_0\dots e_{\mathsf{a}}$ and $e'_{-\mathsf{m}}e'_{-\mathsf{m}+1}\dots e'_0\dots e'_{\mathsf{a}}$ have the same output labeling, then the input labels of the edges e_0 and e'_0 must agree. Such an encoder is called *sliding-block decodable.* It can be decoded via a *sliding-block decoder*, which makes a decision on a given received q-codeword on the basis of the local context of that codeword in the received sequence: the codeword itself, as well as a fixed number m of preceding codewords and a fixed number a of upcoming codewords. In this way, any symbol error at the decoder input should give rise to a limited number of bit errors at the decoder output. Figure 6 shows a schematic diagram of a sliding-block decoder. It is easy to see that a single error at the input to a sliding-block decoder can only affect the decoding of codewords that fall in a "window" of length at most $\mathsf{m}+\mathsf{a}+1$ codewords. We mention that, in symbolic dynamics, any mapping of this type is called a *sliding block code*; a sliding block decoder is a sliding block code which just happens to invert a finite-state encoder.

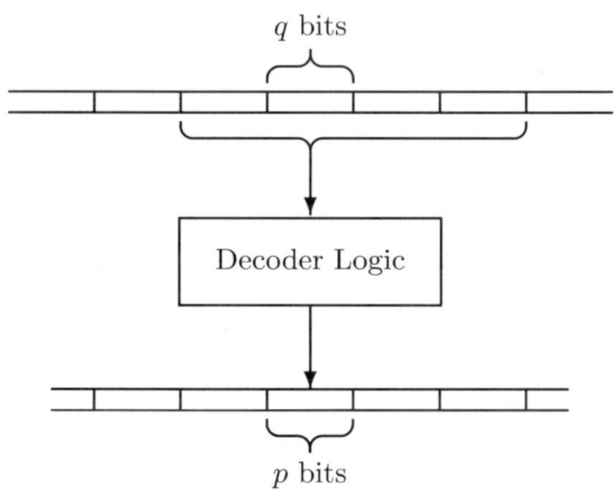

FIGURE 6. Sliding-block decoder schematic.

(current codeword)	(next codeword)	(decoded input)
010	—	11
011	101 or 111	01
011	010, 011, or 110	00
101	101 or 111	10
101	010, 011, or 110	00
110	—	10
111	101 or 111	11
111	010, 011, or 110	01

TABLE 1. Sliding-block decoder for encoder in Figure 5.

EXAMPLE 2.2. Table 1 defines a sliding-block decoder for the encoder in Figure 5. Entries marked by "—" in the table do not affect the value of the decoded input label. For example, according to Table 1, the codeword 010 decodes to 11; the codeword 011 decodes to 01 if it is followed by 101 or 111, and it decodes to 00 if it is followed by 010, 011 or 110.

For a given sofic shift S, we would like to encode user data into S at as high a rate p/q as possible. The highest possible rate is determined by the *entropy* of the sofic shift S, defined as:

$$h(S) = \lim_{n \to \infty} (1/n) \log_2(B_n(S))$$

where $B_n(S)$ is the number of allowed n-blocks in S. The term entropy is often called *capacity*, especially in the engineering literature. In fact, entropy was originally developed by Shannon [**Sha48**], and he called it capacity because he reserved the use of the term entropy for a related probabilistic concept. Later, ergodic theorists borrowed Shannon's entropy, and then developed a topological notion of entropy for general dyanmical systems; for sofic shifts this notion of entropy boils down to the definition of entropy that we have given here.

Among the fundamental results in modulation coding are the following.

THEOREM 2.3. (Finite-state coding theorem)*Let S be a sofic shift. Then there exists a rate $p : q$ finite-state encoder for S if and only if $p/q \leq h(S)$.*

The "only if" part of the preceding result was established by Shannon [**Sha48**], and he also established a weak version of the "if" part. The "if" part, as we have stated it here, is due to Adler, Coppersmith, and Hassner [**ACH83**], who also proved the following stronger version for shifts of finite type:

THEOREM 2.4. (Adler, Coppersmith, and Hassner [**ACH83**]) *Let S be a shift of finite type. If $p/q \leq h(S)$, then there exists a rate $p : q$ sliding-block decodable finite-state encoder into S.*

Both of these results are obtained using an explicit algorithm, called the *state splitting algorithm* or *ACH algorithm*, based on state splitting of graphs; roughly speaking, a *state splitting* of a graph creates a new graph by splitting states of the original graph, with outgoing edges partitioned and incoming edges replicated. State splittings were introduced into symbolic dynamics in order to solve the conjugacy problem [**Wil173**]: given two sofic shifts X and Y, determine when there is an invertible sliding block code (a.k.a. *conjugacy*) from X to Y. It turns out that any conjugacy can be decomposed into a sequence of state splitting and state amalgamation operations.

To prove the results above, one starts with a right resolving presentation (G, L) of S and forms its *q-th power graph* $(H, M) = (G^q, L^q)$, which is the graph consisting of states of G and one labeled edge for each path of length q in G. Then, using the entropy condition $p/q \leq h(S)$, one forms a sequence of state splittings beginning with (H, M) and ending with a graph that has at least 2^p outgoing edges from each state; such a graph can be used as the desired encoder. The choices of state splittings are guided by an *approximate eigenvector*, which is a nonnegative integer vector x that satisfies the vector inequality:

$$A^q x \geq 2^p x; \tag{2.1}$$

(here, A is the adjacency matrix of G, i.e. A_{IJ} is the number of edges from state I to state J). Each state I of G is ultimately split into x_I states to form the encoder.

Theorem 2.4 turns out not to hold in the full generality of sofic shifts, but it can be extended to a class of sofic shifts that is much larger than the shifts of finite type.

3. Error-Correction Codes

3.1. The very basics. In this section, we quickly review some of the basics of error-correction codes (ECC). For a thorough treatment, the reader may consult any one of the excellent textbooks, such as [**Wi95**], on the subject.

An (n, M) *block code* over a finite alphabet F is simply a set of M words each of length n (and so is a subset of F^n of size M). Note that the term "block code" simply refers to a set of words (which are often called *codewords*) and does not imply any specific kind of encoder or decoder.

The *Hamming distance* between two words $\mathbf{x}, \mathbf{y} \in F^n$ is the number of coordinates in which they differ. We denote the Hamming distance by $\mathsf{d}(\mathbf{x}, \mathbf{y})$. It is easy to verify that Hamming distance is indeed a metric. The *minimum distance* of a code $\mathcal{C}$ is the minimum Hamming distance between any two distinct codewords of $\mathcal{C}$. An (n, M) code with minimum distance d is called an *(n, M, d) code*.

Given an (n, M, d) code $\mathcal{C}$ over F, let $\mathbf{c} \in \mathcal{C}$ be a codeword transmitted over a noisy channel, and let $\mathbf{y} \in F^n$ be the received word (so, some values in the codeword may be corrupted by the channel). The number of errors equals $\mathsf{d}(\mathbf{y}, \mathbf{c})$, and the error locations are the indices of the entries in which $\mathbf{c}$ and $\mathbf{y}$ differ. The task of error correction is, of course, to recover $\mathbf{c}$ from $\mathbf{y}$ by recovering the error locations and the error values (in these locations).

The following simple result, which follows directly from the fact that the Hamming distance is a metric, explains how error correction codes can actually correct errors.

PROPOSITION 3.1. Let $\mathcal{C}$ be a block code over F with minimum distance d. For a received word $\mathbf{y}$, there is at most one codeword $\mathbf{c} \in \mathcal{C}$ such that $\mathsf{d}(\mathbf{y}, \mathbf{c}) \leq (d-1)/2$. Thus, if a codeword $\mathbf{c}$ is transmitted over a noisy channel and at most $\lfloor (d-1)/2 \rfloor$ errors are made, then the codeword $\mathbf{c}$ can be recovered from the received word $\mathbf{y}$.

So, for a code of length n and minimum distance $d = 3$, if a channel makes only a single error within a codeword of length n, then the error can be corrected.

3.2. Linear Block Codes. Most of the theory of ECC has focused on linear block codes. Such a code is defined as a finite-dimensional vector space over a finite field $F = \text{GF}(q)$. Typically q is a power of 2, for instance $q = 256$, so that the field elements can represent all possible values of an (8-bit) byte.

The *dimension* of a linear (n, M, d) code $\mathcal{C}$ over F is the dimension of $\mathcal{C}$ as a linear sub-space of F^n. If k is the dimension of $\mathcal{C}$, then we say that $\mathcal{C}$ is a linear $[n, k, d]$ code over F. Note that every basis of a linear $[n, k, d]$ code $\mathcal{C}$ over $F = \text{GF}(q)$ contains k codewords, the linear combinations of which are distinct and generate the whole set $\mathcal{C}$. Therefore, $|\mathcal{C}| = M = q^k$.

Just as for modulation codes, once the desired set of codewords is fixed, one needs to find a way of encoding user words into these codewords. For a general ECC, this can be a difficult problem, but for linear ECC's there is a simple algebraic method, described as follows.

A *generator matrix* of a linear $[n, k, d]$ code over F is a $k \times n$ matrix whose rows form a basis of the code. Let $\mathcal{C}$ be a linear $[n, k, d]$ code over F and G be a generator matrix of $\mathcal{C}$. We can encode user words into codewords of $\mathcal{C}$ by regarding the former as vectors $\mathbf{u} \in F^k$ and encoding $F^k \to \mathcal{C}$ via:

$$\mathbf{u} \mapsto \mathbf{u}G \ .$$

Note that this defines a block encoder of rate $k : n$, which is the maximal rate of any encoder for an $[n, k, d]$ code. The inverse of this encoder mapping is naturally the decoder mapping, which is of course also linear.

Since $\text{rank}(G) = k$, we can apply elementary operations to the rows of G to obtain a $k \times k$ identity matrix as a sub-matrix of G. A $k \times n$ generator matrix is called *systematic* if it has the form

$$(\ I \mid A\ )\ ,$$

where I is a $k \times k$ identity matrix and A is a $k \times (n-k)$ matrix. While a code $\mathcal{C}$ need not have a systematic generator matrix, we can always permute the code coordinates to obtain a code $\hat{\mathcal{C}}$ for which the first k columns of any generator matrix are linearly independent, in which case $\hat{\mathcal{C}}$ has a systematic generator matrix. The code $\hat{\mathcal{C}}$ has the same length, dimension, and minimum distance as the original code $\mathcal{C}$. Often $\hat{\mathcal{C}}$ and $\mathcal{C}$ are regarded as essentially the same code.

When using a systematic generator matrix $G = (\,I \mid A\,)$ for encoding, the mapping $\mathbf{u} \mapsto \mathbf{u}G$ takes the form $\mathbf{u} \mapsto (\,\mathbf{u} \mid \mathbf{u}A\,)$; that is, the information vector is the first part of the codeword; the remainder of the codeword consists of "redundant" symbols; for this reason, for any $[n, k, d]$ linear code $\mathcal{C}$, the difference $n - k$ is called the *redundancy* of $\mathcal{C}$. We will see later that a systematic encoding has a special advantage when concatenating modulation encoders with ECC encoders.

While any linear ECC can be described by a generator matrix, there is an alternative description that is often very convenient. Let $\mathcal{C}$ be a linear $[n, k, d]$ code over F. A *parity-check matrix* of $\mathcal{C}$ is an $r \times n$ matrix H over F such that for every $\mathbf{c} \in F^n$,

$$\mathbf{c} \in \mathcal{C} \qquad \Longleftrightarrow \qquad H\mathbf{c}^T = \mathbf{0}\ .$$

In other words, the code $\mathcal{C}$ is the (right) kernel, $\ker(H)$, of H in F^n. We thus have

$$\text{rank}(H) = n - \dim \ker(H) = n - k\ .$$

The parity check matrix is a useful tool for computing the minimum distance of a linear code:

PROPOSITION 3.2. Let H be a parity-check matrix of a linear code $\mathcal{C} \neq \{\mathbf{0}\}$. The minimum distance of $\mathcal{C}$ is the largest integer d such that every set of $d-1$ columns in H is linearly independent.

The parity check matrix and generator matrix are dual descriptions of a linear block code. One can form a parity-check matrix from a generator matrix and vice versa using simple linear algebra operations. In the special case where G is a systematic matrix $(\, I \,|\, A \,)$, we can take the $(n-k) \times n$ matrix $H = (\, -A^T \,|\, I \,)$ as a parity-check matrix. In this way, the redundant symbols mentioned above in the discussion of a systematic generator matrix may be viewed as parity symbols.

Perhaps the most famous ECC is the linear Hamming code over GF(2), defined by the 3×7 parity-check matrix

$$H = \begin{pmatrix} 0 & 0 & 0 & 1 & 1 & 1 & 1 \\ 0 & 1 & 1 & 0 & 0 & 1 & 1 \\ 1 & 0 & 1 & 0 & 1 & 0 & 1 \end{pmatrix} .$$

The codewords are all of length $n = 7$. Since the rows of H are linearly independent, it follows that the dimension of $\mathcal{C}$ is $k = n - rank(H) = 7 - 3 = 4$. Note that each non-zero vector of length 3 occurs exactly once as a column of H. It follows that the maximum size of a linearly independent set of columns is 2, and so, according to Proposition 3.2, the minimum distance is $d = 3$. So, the Hamming code is a [7,4,3] linear code. As such, we can encode user bits at a rate of 4:7, and according to Proposition 3.1, this code is capable of correcting a single error occurring within each 7-bit codeword.

3.3. The Singleton bound and Reed-Solomon codes. Recall that a good ECC is one in which there are lots of codewords (so, M should be large) and any pair of distinct codewords should differ in many positions (so, d should be large). But for a fixed alphabet size and a fixed codeword length n, these two requirements are at odds with one another. Thus there must be a tradeoff between M and d. The optimal tradeoff is described in the following result.

THEOREM 3.3. (The Singleton bound) *For any (n, M, d) code over an alphabet of size q,*

$$d \leq n - \lceil \log_q M \rceil + 1 .$$

Note that for a linear $[n, k, d]$ code over GF(q) the Singleton bound becomes

$$d \leq n - k + 1 .$$

A code is called *maximum distance separable (MDS)* if it attains the Singleton bound with equality. Note that a linear MDS code can correct a certain number, e, of symbols in error within each codeword if its redundancy, $n-k$, satisfies $n-k = 2e$; in other words, for a linear MDS code, two bytes of redundancy are sufficient (and in fact necessary by the Singleton bound) to correct each error.

The following family of MDS codes is among the most widely used error-correction codes today.

Let $\alpha_1, \alpha_2, \ldots, \alpha_n$ be distinct nonzero elements of $F = \mathrm{GF}(q)$. A *Reed-Solomon code* over F is a linear $[n, k, d]$ code with the parity-check matrix

$$H = \begin{pmatrix} 1 & 1 & \ldots & 1 \\ \alpha_1 & \alpha_2 & \ldots & \alpha_n \\ \alpha_1^2 & \alpha_2^2 & \ldots & \alpha_n^2 \\ \vdots & \vdots & \vdots & \vdots \\ \alpha_1^{n-k-1} & \alpha_2^{n-k-1} & \ldots & \alpha_n^{n-k-1} \end{pmatrix}.$$

Note that this construction requires an upper bound on codeword length in terms of alphabet size: $n < q$.

PROPOSITION 3.4. Every Reed-Solomon code is MDS.

This result follows from Proposition 3.2 and the Vandermonde form of H.

3.4. Soft Information for Decoding. For a linear block code $\mathcal{C}$, we found that a generator matrix provides a simple way to construct an encoder. The inverse mapping is of course the decoder. But typically in the theory (and practice!) of ECC, decoding means something quite different from inverting the encoder: it usually means a method of finding the transmitted codeword given the received word. As illustrated in section 3.1, if a code has minimum distance 3, and at most one error occurs in any codeword, then the correctly transmitted codeword can be recovered. Once the correct codeword is found, the user information is recovered by applying the inverse of the encoder. In summary, the decoding process for an ECC contains two qualitatively different steps: finding the right codeword $\mathbf{c}$ and then translating $\mathbf{c}$ to the user information word.

All of this assumes that channel errors are made by flipping 0's to 1's or 1's to 0's. But in many practical channels, the transmitted or recorded signals have continuous values. If the bits in a codeword are supposed to carry nominal real values, say r_0 for 0 and r_1 for 1, then the channel may cause corruption by adding a "noisy" real number r_n to the nominal value. So, if the received value at a bit position is r, how can we tell whether a 0 or 1 was actually transmitted? One way to do this would be to simply see whether r is closer to r_0 or r_1. However, this "bit-by-bit decoding" does not exploit correlations among the bit values within the codewords; so, it may be useful to pass the received values r for all bit locations in the codeword to another processor that may make a better guess of the codeword that was actually transmitted. This information is called *soft information* and is often used to improve the decoding process.

Soft information is critical for the successful decoding of a recently developed class of codes, known as turbo codes and the closely related, but much older, low density parity check codes [**Mac99**]. Moreover, while the very powerful Reed-Solomon codes have traditionally been decoded via very efficient hard decoders, soft decoding algorithms have recently been developed for these codes as well [**KV01**].

4. Concatenation of Encoders

Ideally, user data would be encoded by a single encoder that generates a set of sequences satisfying both ECC and modulation properties. While there has been a good deal of work done on this, it is not so easy to find simple sets of sequences that have both properties and also have enough entropy to yield reasonably high rate codes. So instead, most recording systems employ two encoders, one for ECC and the other for modulation. In *standard concatenation*, illustrated in Figure 7, messages are first passed through an error-correction encoder and then a modulation encoder before being transmitted across the channel. At the receiver, data is decoded via the modulation decoder and then the error-correction decoder. Standard concatenation seems quite natural because the modulation encoder produces sequences that must satisfy constraints dictated by the channel; otherwise, the error-correction encoder might well destroy the modulation properties.

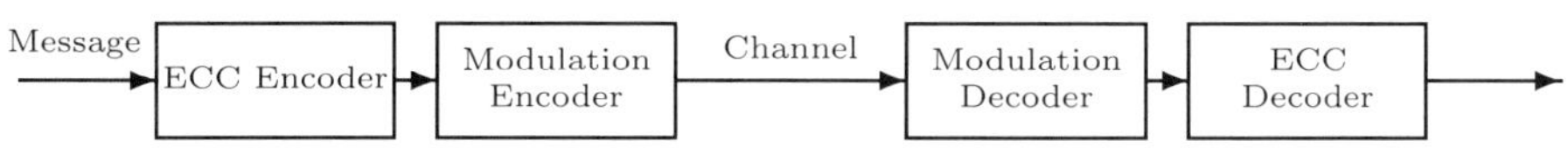

FIGURE 7. Standard concatenation

If the ECC decoder requires soft information, as we have discussed in section 3.4, then this information must be filtered through the modulation decoder (before it arrives at the ECC decoder). This process can be complicated and introduce inaccuracies.

Now, suppose that somehow the modulation encoder could produce sequences that not only satisfy the modulation constraints, but also whenever the bit values in certain positions (called *unconstrained positions*) are flipped (or not flipped) independently, the modulation constraints are still not violated. ECC parity information, obtained from a systematic ECC encoding of the modulation encoded stream, can then be inserted into these unconstrained positions without violating the constraint. This scheme, called the *parity insertion scheme* is illustrated in Figure 8. The modulation encoder produces a sequence $\mathbf{w}$, which is input to a systematic ECC encoder that produces parity bits $\mathbf{r}$; then the parity bits are inserted into the unconstrained positions, without fear of violating the constraint, and passed through the channel. This scheme was introduced in [**WI98**]; see also [**WI01**], [**Imm99**, pp.103-105], [**F99**], [**FCi99**].

As a simple example of a modulation encoder which allows certain bits to be unconstrained, consider the length-5 block code [**WI01**] for the RLL(0,2) shift:

$$\{01010, 10010\}.$$

Note that indeed all concatenations of these words satisfy the RLL(0,2) shift; moreover, this still holds if the bit values in the third and fifth positions are flipped independently. We can view this as a rate 1:5 modulation encoder with two bit positions reserved for ECC parity. Of course, this works because the original block code above satisfies a much stronger constraint than is necessary to satisfy RLL(0,2); this leaves "extra room" for some positions to be unconstrained.

The general subject of constructing modulation encoders that produce sequences with unconstrained bit positions is taken up in section 5.

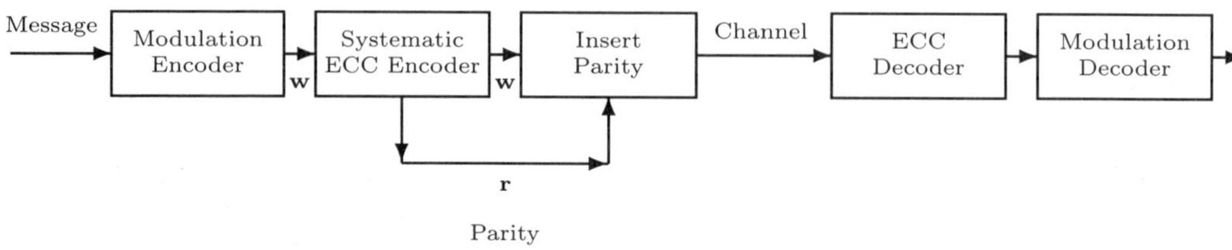

FIGURE 8. Parity Insertion

5. Constrained systems with unconstrained positions

Let S be a sofic shift, N a positive integer and $U \subseteq \{0, \ldots, N-1\}$ (the notation "U" is supposed to suggest "Unconstrained", and U will sometimes be called the *unconstrained set*). We say that a sequence x' is a *U-flip* of a sequence x if

$$x'_i = x_i \text{ whenever } i \bmod N \notin U.$$

In other words, x' is obtained from x by independently flipping (or not flipping) the bit values in positions:

$$\{i \in Z : \ i \bmod N \in U\}.$$

The *(U, N)-unconstrained version of S*, denoted $S_{U,N}$, is the set of all shifts of all sequences $x \in S$ such that:

(1) $x_i = 0$ for all $i \bmod N \in U$ – and –
(2) all U-flips of x belong to S.

Note that in the unconstrained positions, the bit value is forced to be "0", but this was arbitrary: we could have just as well chosen "1" or made a random, but fixed, assignment of bit values in these positions. These positions are unconstrained in the sense that we can independently change the bit values without violating the constraint.

The ratio $|U|/N$ is called the *parity insertion rate* (or simply *insertion rate*) because it represents the percentage of positions in which ECC parity information can be inserted without violating the constraint.

In the following, we show that $S_{U,N}$ is again a sofic shift, by exhibiting a presentation of it. For this, we need the notion of follower sets from symbolic dynamics. Given a sofic shift S and a finite sequence u, the *follower set* of u is:

$$F(u) = F_S(u) = \{\text{finite sequences } v : uv \text{ is allowed in } S\}.$$

It is well known that any sofic shift has only finitely many follower sets [**LM95**], [**MRS98**].

Consider the following labeled graph $G_{U,N}$.

- **States:** All pairs of the form (ℓ, F) where $\ell \in \{0, \ldots, N-1\}$ and F is an intersection (possibly empty) of one or more follower sets of S
- **Edges:** For $\ell \notin U$, we have the edges:

$$(\ell,\ F(u_1) \cap \cdots \cap F(u_k)) \xrightarrow{a} (\ell + 1 \bmod N,\ F(u_1 a) \cap \cdots \cap F(u_k a))$$

provided that a belongs to $F(u_1) \cap \cdots \cap F(u_k)$. For $\ell \in U$, we have the edges:

$$(\ell,\ F(u_1) \cap \cdots \cap F(u_k))$$
$$\xrightarrow{0} (\ell + 1 \bmod N,\ F(u_1 0) \cap \cdots \cap F(u_k 0) \cap F(u_1 1) \cap \cdots \cap F(u_k 1))$$

provided that both 0 and 1 belong to $F(u_1) \cap \cdots \cap F(u_k)$.

Note that $G_{U,N}$ is periodic with period N in the sense that the states of $G_{U,N}$ naturally divide into N phases, with edges cyclically leading from one phase to the next.

Of course, one must verify that the edges that specify $G_{U,N}$ are well-defined (indeed, they are!). With that, the following result is not hard to prove.

PROPOSITION 5.1. $G_{U,N}$ is a presentation of $S_{U,N}$, and so $S_{U,N}$ is a sofic shift.

As an example, for the RLL$(0,k)$ shift, $G_{U,N}$ is as follows.

- **States:** $k+1$ states in each of N phases:
$$\{(\ell, i) : 0 \le \ell \le N-1,\ 0 \le i \le k\}$$
here i represents the number of preceding consecutive 0's and ℓ represents the phase;
- **Edges:** beginning in any phase $\ell \notin U$, we have the edges:
$$(\ell, i) \xrightarrow{0} (\ell + 1 \bmod N, i+1) \quad \text{if } i < k$$
$$(\ell, i) \xrightarrow{1} (\ell + 1 \bmod N, 0)$$
and beginning in any phase $\ell \in U$ we have the edges:
$$(\ell, i) \xrightarrow{0} (\ell + 1 \bmod N, i+1) \quad \text{if } i < k$$

For RLL(0,4), let $N = 3$ and $U = \{1\}$; then $G_{U,N}$ is illustrated in Figure 10 (with the phase 0 states repeated at both top and bottom). In any phase $\ell \notin U$, the edges mimic those of the standard presentation of RLL(0,4) (as shown in Figure 9), but pass from phase ℓ to phase $\ell + 1$. In any phase $\ell \in U$, the next binary symbol is constrained to be 0, but must allow for the possibility that the 0 can be flipped to a 1 without violating the constraint; thus, the edge outgoing from state (ℓ, i) must end at the "more severely constrained" of the two possibilities: $(\ell+1, i+1)$ or $(\ell+1, 0)$, namely $(\ell+1, i+1)$ (provided of course that $i < k$; otherwise, there is no edge).

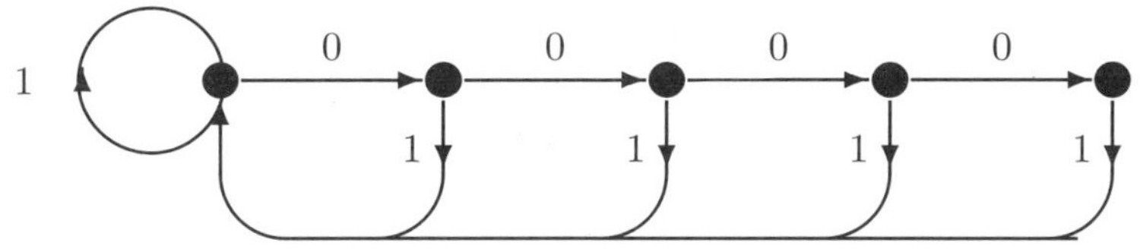

FIGURE 9. RLL(0,4)

Given a sofic shift S, a positive integer N and a specified set U of positions modulo N, we can apply the ACH Algorithm to $G_{U,N}$, as discussed in section 2.3, to

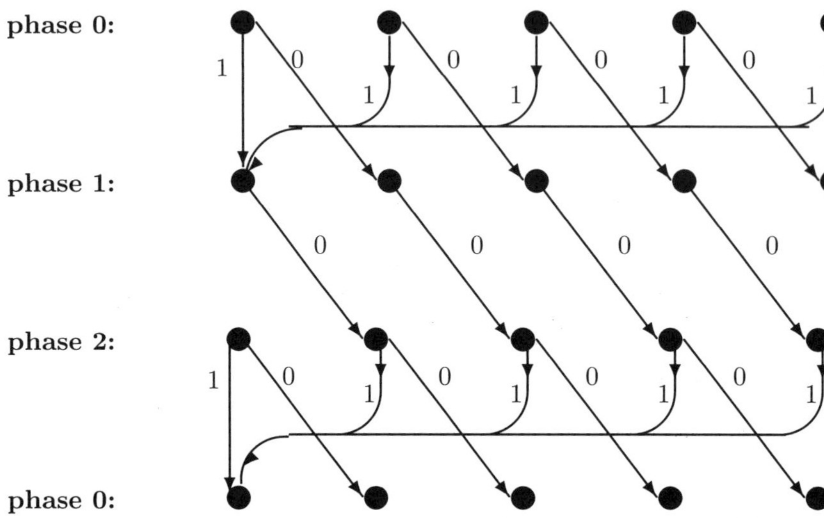

FIGURE 10. $G_{U,N}$ for RLL$(0,4)$, $N = 3$ and $U = \{1\}$

obtain finite-state encoders for $S_{U,N}$. These tend to have better rate and decoding properties than simple block encoders.

When N is large, $G_{U,N}$ will necessarily have many states. However, an encoder for $S_{U,N}$ need not use all N phases. For instance, a rate $p : N$ encoder need use only one phase. In general, a rate $p : q$ encoder need use only

$$\frac{N}{\text{g.c.d}(N,q)} \tag{5.1}$$

phases of $G_{U,N}$; each of these encoder phases can be viewed as a rate $p : q$ finite-state machine with initial states in one phase and terminal states in another phase [**AsMa00**].

Even if S is a shift of finite type, it turns out that $S_{U,N}$ need not have finite type. So, Theorem 2.4 does not guarantee sliding block decodability. However, the ACH algorithm (which implements the proof of that result) applies to the systems $S_{U,N}$ to provide finite-state encoders that are sliding block decodable, *provided that the decoder has phase information available* (i.e., knowledge of the phase mod N). At the end of the ACH algorithm encoder states that have been split from parents that were not in the same phase may yield inconsistent decoding functions; so, the decoding function may vary from phase to phase. Since phase information is often readily available to the decoder, this is not much of a problem.

Table 2 presents a list of four rate 8/9 encoders for RLL $(0,4)$. Encoder #1 is a block encoder with the shortest block length q that permits a nonzero parity insertion rate and code rate at least 8/9 (a standard code rate in the recording industry). Here, $q = N = 18$ and the parity insertion rate is $1/18 \approx .0556$ (the encoder operates at rate 16:18). It turns out that if one wants to strictly increase the insertion rate, but still keep the code rate at least 8/9, then for a block encoder, the block length must increase to $q = N = 45$. This can be achieved by a block encoder with three unconstrained bit positions, and so the insertion rate is $3/45 =$

$1/15 \approx .0667$ (shown as encoder #2). However, the large block length means that encoding will probably be very complicated. On the other hand, the same code rate (8/9) and and same insertion rate (1/15) can be achieved via a sliding-block decodable finite-state encoder (encoder #3) with block length only $q = 9$ (here, $N = 15$ and there is one unconstrained position). Moreover, the decoder window has length = two 9-bit blocks, and so is comparable (in number of bits) to encoder #1 (and much shorter than that of encoder #2). Also, according to (5.1) only 5 of the 15 phases need be used for this encoder, and it turns out that the ACH algorithm yields such an encoder with only about 5 states per phase. Finally, encoder #4 (also constructed using the ACH algorithm) again has block length $q = 9$, code rate $= 8/9$, and further improved insertion rate $2/27 \approx .0741$. But this encoder requires requires a larger decoding window (three 9-bit blocks).

scheme	*length*	*code rate*	*parity insertion rate*	*window (bits)*	*#states*
#1 *block*	18	8/9	$1/18 \approx .0556$	18	1
#2 *block*	45	8/9	$3/45 \approx .0667$	45	1
#3 *finite state(ACH)*	9	8/9	$1/15 \approx .0667$	18	26
#4 *finite state(ACH)*	9	8/9	$2/27 \approx .0741$	27	32

TABLE 2. Rate 8/9 encoders for RLL(0, 4)

Much more detail on the subject of constrained systems with unconstrained positions can be found in [**CMNW01**]. That paper discusses in greater depth the tradeoffs between code rate and parity insertion rate, in particular the optimal tradeoff as the block length grows.

6. Reversed Concatenation with Lossless Compression

Standard concatenation, as illustrated in Figure 7, has the disadvantage that the modulation decoder, which comes before the error-correction decoder, can propagate channel errors before they can be corrected. Of course, as discussed in section 2.3, we require the modulation decoder to be sliding-block decodable in order to limit error propagation. But if the rate $p : q$ of the modulation encoder is very close to the entropy of the sofic shift, then q will typically need to be very large, and so even if the sliding block window were only one block wide, the extent of error propagation may be unacceptable.

The parity insertion scheme described in section 4 is a particular type of *reversed concatenation* scheme. But usually reversed concatenation means something like what is illustrated in Figure 11 [**Bli81**], [**Man91**], [**FCa98**], [**Imm97**]. Here is the general idea.

First, user data $\mathbf{u}$ is encoded via the *data modulation encoder* E1, producing a sequence $\mathbf{w}$. This encoder may have very high rate based on very long block length q. Then $\mathbf{w}$ is passed through a systematic ECC encoder, which outputs $\mathbf{w}$ verbatim as well as a block $\mathbf{r}$ of (redundant) parity bits. Now, $\mathbf{w}$ satisfies the modulation constraints, but $\mathbf{r}$ need not. So, $\mathbf{r}$ is encoded to a sequence $\mathbf{y}$ via a second modulation encoder E2, called the *parity modulation encoder*. Then $\mathbf{y}$ is appended to the modulation-encoded data stream $\mathbf{w}$ and transmitted across the channel. Some arrangement needs to be made to ensure that the entire stream satisfies the modulation constraints. For an RLL constraint, typically a few merging bits are inserted between $\mathbf{w}$ and $\mathbf{y}$ for this purpose.

After passing through the channel, corresponding sequences $\hat{\mathbf{w}}$ and $\hat{\mathbf{y}}$ are separated. The latter is then decoded via the parity modulation decoder D2, recovering $\hat{\mathbf{r}}$, a possibly corrupted version of the ECC-parity $\mathbf{r}$. Here, it is essential for D2 to limit error propagation very tightly (so, while E1 may be a very long-block-length encoder, the design of E2 must pay attention to error-propagation, such as in the ACH algorithm). In this way, many channel errors can be corrected before the data modulation decoder, D1, thereby mitigating the problem of error propagation associated with the long block length of E1. Moreover, if the data modulation encoder and ECC encoder have high code rate, then the overall scheme will still have high code rate because the length of the parity stream, $\mathbf{r}$, relative to $\mathbf{w}$, will be rather small.

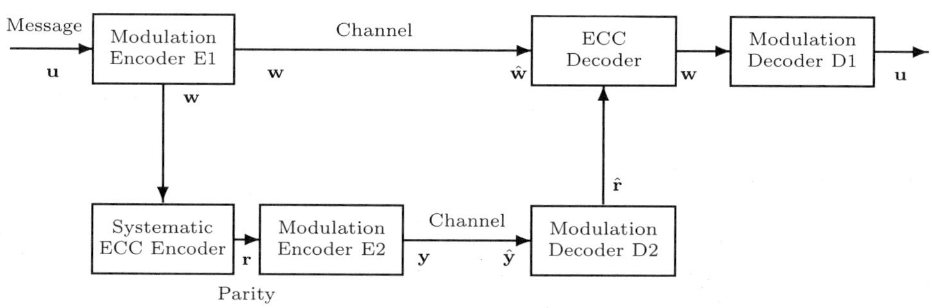

FIGURE 11. Reversed concatenation

Reversed concatenation, as illustrated in Figure 11, suffers one particular disadvantage compared with standard concatenation, as illustrated in Figure 7. Namely, a burst of channel errors in standard concatenation will naturally be compressed by the modulation decoder, thereby lessening some burden on the ECC decoder. So, in reversed concatenation, the ECC decoder will need to correct bursts of longer lengths than in standard concatenation; this poses an extra burden on the ECC.

However, the performance of reversed concatenation can be improved if one *compresses* the data modulation encoded sequence $\mathbf{w}$ in a lossless (one-to-one) manner before being passed on to the ECC encoder [**Imm97**]. In Figure 12, $\mathbf{w}$ is compressed into a sequence $\mathbf{s}$, which is fed to the ECC encoder, which then generates a sequence of parity symbols $\mathbf{r}$ based on $\mathbf{s}$. Since $\mathbf{s}$ is shorter than $\mathbf{w}$, one might expect that compression schemes would have the effect of shortening burst errors that occur in in $\mathbf{w}$, thereby lowering the redundancy of the ECC scheme. At the channel output, the received sequence $\hat{\mathbf{w}}$ is compressed to a string $\hat{\mathbf{s}}$, and the ECC decoder recovers $\mathbf{s}$ from $\hat{\mathbf{r}}$ and $\hat{\mathbf{s}}$. Then the decompressor recovers $\mathbf{w}$, and the modulation decoder D1 recovers $\mathbf{u}$.

At one extreme, we could compress $\mathbf{w}$ back to $\mathbf{u}$ (in which case $\mathbf{s}$ would be the same as $\mathbf{u}$). But then a small channel error in $\hat{\mathbf{w}}$ could corrupt all of $\hat{\mathbf{s}}$ *before* error correction. Instead, the compression scheme will guarantee that such a channel error can corrupt only a limited number of symbols in $\hat{\mathbf{s}}$. For this purpose, it is useful to employ sliding block compressors, discussed in section 7.

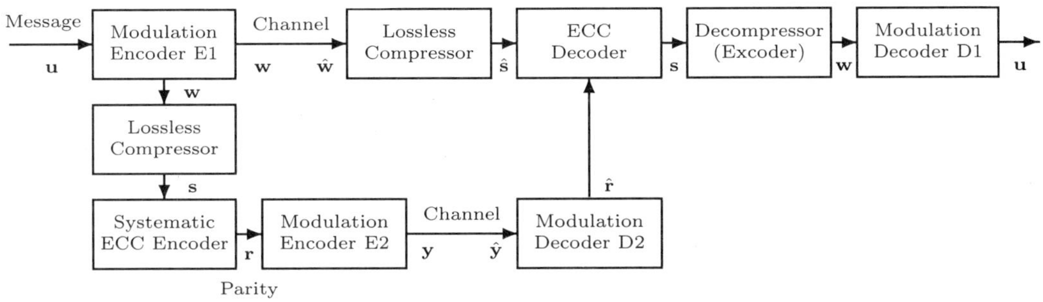

FIGURE 12. Reversed concatenation with lossless compression.

7. Sliding block compressors

A lossless compression scheme will consist of a *compressor* and an *expanding coder* (in short, *excoder*), which acts as a "decompressor". We define this formally as follows.

Let S be a sofic shift and n a positive integer. A *sliding-block compressible finite-state* (S, n)*-excoder* is a graph $\mathcal{E}$ which has two labelings: an input labeling I over the alphabet $\{0, \ldots, n-1\}$ and an output labeling O over the alphabet of S such that:

(1) $(\mathcal{E}, I)$ is right resolving (i.e., at each each state, the outgoing edges are assigned distinct input labels); in particular, each state has at most n outgoing edges.
(2) every sequence in S is generated by $(\mathcal{E}, O)$
(3) there are integers m, a such that if $\mathbf{w}$ is a word allowed in S and two paths $e_{-\mathsf{m}}e_{-\mathsf{m}+1}\cdots e_0\cdots e_{\mathsf{a}}$ and $e'_{-\mathsf{m}}e'_{-\mathsf{m}+1}\cdots e'_0\cdots e'_{\mathsf{a}}$ in $\mathcal{E}$ are both O-labeled by $\mathbf{w}$, then the edges e_0 and e'_0 have the same I-label.

The definition of sliding-block compressible excoder is dual to that of sliding-block decodable encoder: the main difference is in the containment relationship between the given sofic shift S and the sofic shift presented by $\mathcal{E}$: any sequence generated by an eNcoder must satisfy the constraint; here, any sequence that satisfies the constraint must be generated by the eXcoder. Note that, while $(\mathcal{E}, O)$ may generate sequences that violate the constraint, condition (3) applies only to those paths in $\mathcal{E}$ that generate words allowed in S.

A rate $p : q$ sliding-block compressible excoder for S is defined to be a sliding-block compressible $(S^q, 2^p)$-excoder. The compressor is a sliding-block code from sequences of q-blocks of a sofic shift S to unconstrained sequences of p-blocks over $\{0, 1\}$. The excoder, on the other hand, has the form of a finite-state machine, similar to encoders.

A *rate* $p : q$ *block excoder for* S is a rate $p : q$ excoder for S with one state. The corresponding sliding-block compressor is simply a block code, called a *block compressor*.

The next result is dual to Theorem 2.4 and can be used to construct sliding-block compressible excoders that have some advantages compared with the simpler block excoders.

PROPOSITION 7.1. Let S be a shift of finite-type such that $h(S) \leq p/q$. Then there is a sliding-block compressible rate $p : q$ excoder for S.

The proof of this result is algorithmic and was motivated by the ACH Algorithm discussed in section 2.3; see [**Mar85**]. In particular, the proof of Proposition 7.1 employs a state-splitting algorithm based on a dual notion of approximate eigenvector, which satisfies a reversed inequality compared with inequality (2.1).

The error correction effect on error burst size will depend on several factors, including the compression rate p/q, the sliding block window size and edge effects having to do with how the error burst matches up with the codeword boundaries. Generally, for a given block length q, we can achieve better compression rates with sliding-block compressible excoders than with block excoders; but the edge effects from the sliding window can degrade this advantage. For the sofic shift $\mathrm{RLL}(2, \infty)$, presented in Figure 13, we compare, in Figure 14, three excoders of roughly the same complexity. Two of these are block excoders and one is a sliding-block compressible excoder constructed from the algorithm used to prove Proposition 7.1.

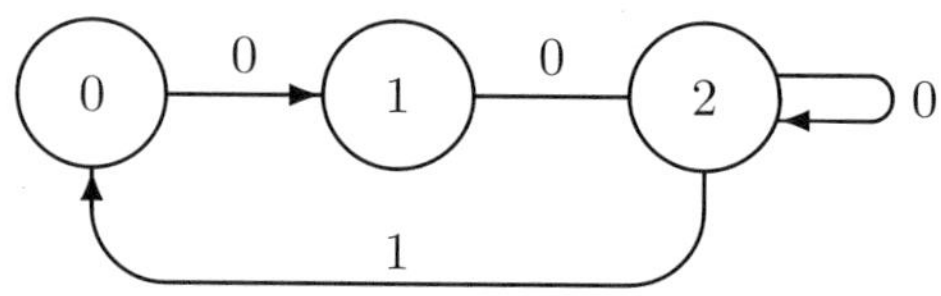

FIGURE 13. Graph presentation $G_{2,\infty}$ of $\mathrm{RLL}(2, \infty)$ shift

In this figure, the horizontal axis represents the length of a single burst of channel errors, and the vertical axis represents the number of parity bits required to correct such a burst in a sector or 4,096 user bits (the standard sector size in today's magnetic hard disk drives). We assumed that the ECC is an interleaved single-error correcting Reed-Solomon code, optimized over n-bit alphabets, $n = 4, \ldots, 8$. The figure indicates that the block excoders have some advantage for short burst length L. However, for $L \geq 80$, the rate 4:7 sliding-block compressible excoder appears to be a winner.

Much more detail on the subject of sliding block compressible excoders can be found in [**FMR00**].

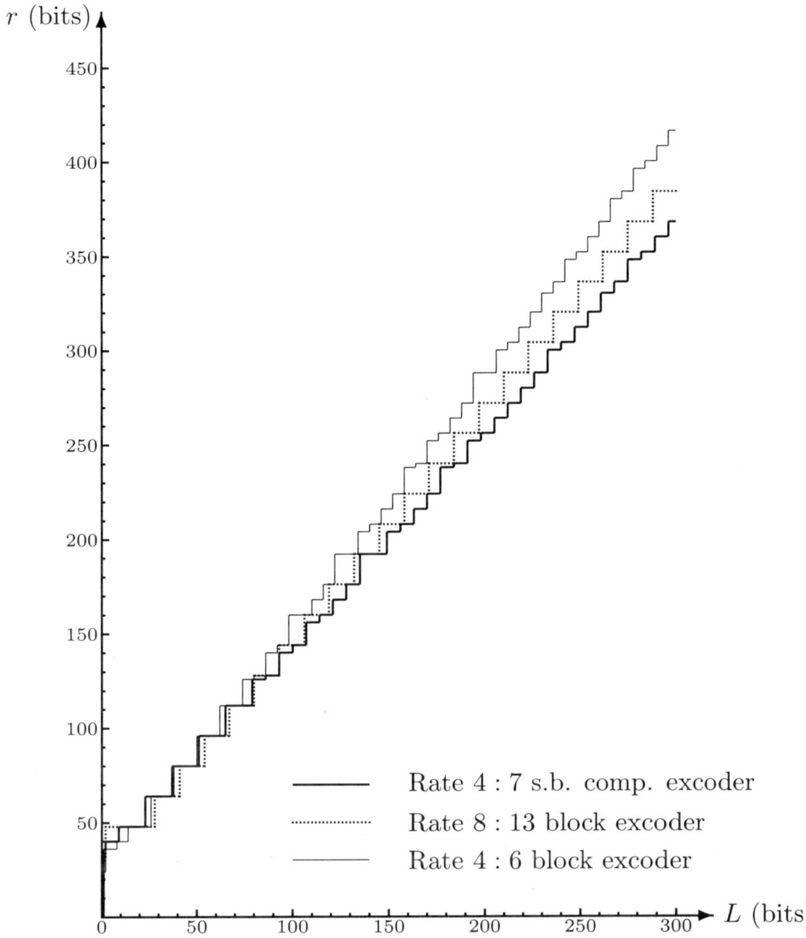

FIGURE 14. Redundancy for 4,096 user bits and various burst lengths using three different excoders for the RLL$(2, \infty)$ shift.

References

[ACH83] R.L. Adler, D. Coppersmith, M. Hassner, *Algorithms for sliding block codes — an application of symbolic dynamics to information theory, IEEE Trans. Inform. Theory,* 29 (1983), 5–22.

[AnMc99] K. Anim-Appiah and S. McLaughlin, *Constrained-input turbo codes for (0,k) RLL channels, CISS Conference,* Baltimore, 1999.

[AnMc00] K. Anim-Appiah and S. McLaughlin, *Towards soft output ASPP decoding for non-systematic nonlinear block codes,* 2000.

[AsMa00] J. Ashley and B. Marcus, *Time-varying encoders for constrained systems: an approach to limiting error propagation, IEEE Trans. Inform. Theory,* 46 (2000), 1038 - 1043.

[Bli81] W.G. Bliss, *Circuitry for performing error correction calculations on baseband encoded data to eliminate error propagation, IBM Tech. Discl. Bull.,* 23 (1981), 4633–4634.

[CMNW01] J. Campello, B.Marcus, R.New, B.Wilson, *Constrained systems with unconstrained positions, IEEE Tranas. Inform. Theory,* 48 (2002), 866-879.

[F99] J. Fan, *Constrained coding and soft iterative decoding for storage,* PhD Dissertation, Stanford University, 1999.

[FCa98] J. Fan and R. Calderbank, *A modified concatenated coding scheme, with applications to magnetic data storage, IEEE Transactions on Information Theory,* 44 (1998), 1565 - 1574.

[FCi99] J. Fan, J. Cioffi, *Constrained coding techniques for soft iterative decoders, IEEE Globecom Proceedings,* 1999.

[FMR00] J. Fan, B. Marcus and R. Roth, *Lossles sliding-block compression of constrained systems, IEEE Transactions on Information Theory,* 46 (2000), 624 - 633.

[Imm97] K.A.S. Immink, *A practical method for approaching the channel capacity of constrained channels, IEEE Trans. Inform. Theory,* 43 (1997), 1389 - 1399.

[Imm99] K.A.S. Immink, *Codes for Mass Data Storage Systems,* Shannon Fouundation Publishers, The Netherlands, 1999.

[Imm01] K.A.S. Immink, *A survey of codes for optical disk recording, IEEE Journal on selected areas in communications,* 19 (2001), 756-764.

[Kit98] B. Kitchens, *Symbolic Dynamics: one-sided, two-sided and countable state Markov shifts,* Springer-Verlag, 1998.

[KV01] R. Koetter, A. Vardy, *Algebraic Soft-Decision Decoding of Reed-Solomon Codes, IEEE Trans. Inform. Theory,* submitted, 2001.

[LM95] D. Lind, B. Marcus, *An Introduction to Symbolic Dynamics and Coding,* Cambridge University Press, 1995 (reprinted 1999).

[Mac99] D.J.C. MacKay, *Good error-correcting codes based on very sparse matrices, IEEE Trans. Inform. Theory,* 45 (1999), 399 - 431.

[Man91] M. Mansuripur, *Enumerative modulation coding with arbitrary constraints and post-modulation error correction coding and data storage systems, Proc. SPIE,* 1499 (1991), 72–86.

[Mar85] B.H. Marcus, *Sofic systems and encoding data, IEEE Trans. Inform. Theory,* 31 (1985), 366–377.

[Mar00] B. H. Marcus, *Modulation codes for holographic recording, Holographic Storage,* in Holographic Storage, ed. H. Coufal, D. Psaltis, G. Sincerbox, Springer-Verlag, 2000, pp. 283-292.

[MRS98] B. Marcus, R. Roth, P. Siegel *Constrained systems and coding for recording channels,,* Chapter 20 of Handbook of Coding Theory, ed. V. Pless, C. Huffman, 1998, Elsevier.

[MB96] J. Moon, B. Brickner, *Maximum transition run codes for data storage, IEEE Trans. Magnetics,* 32, no. 5, 39992–3994, 1996.

[Pat75] A.M. Patel, *Zero-modulation encoding in magnetic recording, IBM J. Res. Develop.,* 19 (1975), 366–378.

[Sha48] C.E. Shannon, *The mathematical theory of communication, Bell Sys. Tech. J.,* 27 (1948), 379–423.

[UB88] C. M. J. van Uijen and C.P.M.J. Baggen, *Performance of a class of channel codes for asymmetric optical recording,* Proc. 7th Int. Conf. on Video and Data Recording, IERE Conference Publication no. 79, (1988), 29-32.

[WT99] S.X. Wang, A. M. Taratorin, *Magnetic Information Storage Technology,* Academic Press, 1999.

[Wi95] S. B. Wicker, *Error Control Coding in Digital Communication and Storage,* Prentice-Hall, 1995

[WF83] A. Widmer and P. Franaszek, *A DC-balanced, partitioned-block 8b/10b transmission code, IBM J. R. & D.,* 27 (1983), 440 – 451.

[WI98] A. Wijngaarden, K. A.S. Immink, *Efficient error control schemes for modulation and synchronization codes, Proceedings of the International Sympossium on Information Theory,* 1998, p. 74.

[WI01] A. Wijngaarden, K. A. S. Immink, *Maximum Run-length Limited Codes with Error Control Properties, IEEE Journal on selected areas in communications,* 19 (2001), 602-611.

[Will73] R.F. Williams, *Classification of subshifts of finite type, Annals Math.,* 98 (1973), 120–153; errata: *Annals Math.,* 99 (1974), 380–381.

IBM Almaden Research Center
E-mail address: marcus@almaden.ibm.com

Proceedings of Symposia in Applied Mathematics
Volume **60**, 2004

Complex Dynamics and Symbolic Dynamics

Paul Blanchard, Robert L. Devaney, and Linda Keen

ABSTRACT. In this paper we describe a remarkable connection between the group of automorphisms of the one-sided d-shift and the topology of the parameter plane for complex polynomial dynamical systems.

1. Introduction

Our goal in this paper is to present an interesting connection between two very different areas of dynamical systems, complex dynamics and symbolic dynamics. Usually, a dynamicist would regard symbolic dynamics as a tool to understand a particular (and usually very complicated) type of dynamical system. In this paper, however, we will take the opposite tack: we will use complex dynamics to explain an interesting topic in symbolic dynamics.

Specifically, we will consider the structure of the parameter space for complex polynomial maps. For a degree $d \geq 2$ polynomial, the parameter space is $\mathbb{C}^{d-1}$, i.e., one complex parameter for each finite critical point of the map. This parameter space divides into two distinct regions. One region corresponds to the set of parameters for which all critical points escape to ∞, the so-called escape locus. The complementary region consists of those parameters for which one or more critical points have bounded orbits. We call this (somewhat inaccurately) the boundedness locus. In the quadratic case, this set of parameters is the well-known Mandelbrot set. Note that, up to homotopy, we have only one non-trivial closed curve in the complement of the Mandelbrot set. In the degree d case, the escape locus is much more complicated from a topological point of view: there are infinitely many topologically distinct curves in this piece of parameter space. That is, the topology of the boundedness locus is much more complicated when the degree is 3 or more.

In complex dynamics, the important object from the dynamics point of view is the Julia set. It is on this set where all of the interesting chaotic dynamics resides. When we choose a parameter from the escape locus, then it is known that the Julia set is homeomorphic to a Cantor set and the dynamics on this set is equivalent

2000 *Mathematics Subject Classification.* 37B10.

Key words and phrases. Complex dynamics, symbolic dynamics, shift map, automorphism of the shift.

This paper is a summary of the talk presented by one of the authors (RLD) at the AMS Short Course on Symbolic Dynamics and its Applications held in San Diego, January 4-5, 2002.

to the one-sided shift map (described below). The shift map is without doubt the fundamental object in symbolic dynamics. Now when we follow this Cantor set as we traverse a closed curve in the escape locus, we induce a map on the Julia set, the monodromy map. If the loop is homotopically trivial, then the induced map on the Cantor set is the identity map. But if the loop is non-trivial, then so is the induced map on the Cantor set. More importantly, this induced map commutes with the shift map. Thus we get what is called an automorphism of the shift. An important problem in symbolic dynamics was to describe the set (actually, group) of all automorphisms of the shift. This was resolved in 1990, thanks to work of Boyle, Franks, and Kitchens [**BFK**] and Ashley [**Ash**]. Our goal in this paper is to show that all automorphisms of the shift arise from such monodromies in the escape locus.

This leads to a much more difficult unsolved question: what about automorphisms of the two-sided (invertible) shift map? It is known that the group structure here is much more complicated than in the one-sided case. At the end of this paper we discuss some musings about this situation, and suggest that it is the Hénon mapping that may shed some dynamical light on the structure of this group.

2. Automorphisms of the Shift

In this section we describe some preliminary material regarding automorphisms of the one-sided d-shift. Let Σ_d denote the space of sequences on the d symbols $0, 1, \ldots, d-1$. We endow Σ_d with the usual topology, i.e., two sequences are close if their first n entries are the same. Let $\sigma : \Sigma_d \to \Sigma_d$ denote the shift map given by

$$\sigma(s_0 s_1 s_2 \ldots) = (s_1 s_2 s_3 \ldots).$$

As is well known, σ is a continuous, d-to-1 map onto Σ_d.

An *automorphism* of the d-shift is a homeomorphism $\phi : \Sigma_d \to \Sigma_d$ that commutes with the shift map, i.e., $\phi \circ \sigma = \sigma \circ \phi$. The collection of all such automorphisms, Aut_d, is clearly a group, and a natural question in symbolic dynamics is to describe the structure of this group. Our goal in this paper is to describe a set of generators for Aut_d.

EXAMPLE 1. Let $\phi : \Sigma_2 \to \Sigma_2$ be the map that interchanges each 0 and 1 in a sequence in Σ_2. Clearly, $\phi \in \mathrm{Aut}_2$. A result of Hedlund [**H**] shows that ϕ is the only non-trivial ($\neq$ identity) automorphism of Σ_2.

EXAMPLE 2. In similar fashion, the map generated by any permutation of the d symbols $0, 1, \ldots, d-1$ generates an automorphism of Σ_d. As mentioned above, Aut_d consists of many other, very different automorphisms when $d > 2$.

EXAMPLE 3. Consider the map ϕ_0 defined on Σ_3 as follows. Given a sequence $s_0 s_1 s_2 \ldots$, if $s_{j+1} = 0$, then the map interchanges 1 and 2 if these digits appear in the s_j slot. If $s_j = 0$, nothing happens. For example

$$\phi_0(10120012100\ldots) = (20110012200\ldots)$$

ϕ_0 is easily seen to be an automorphism of Σ_3. We call the digit 0 a *marker* and the map ϕ_0 a *marker automorphism.*

EXAMPLE 4. Again in Σ_3, we let $\overline{0}$ stand for the digit "not 0," i.e., $\overline{0}$ represents either 1 or 2. Let $\phi_{\overline{0}}$ denote the map of Σ_3 that interchanges a 1 and 2 whenever a 1 or 2 is followed by $\overline{0}$. For example

$$\phi_{\overline{0}}(11200102100\ldots) = (22200101100\ldots)$$

As before, it is easy to check that $\phi_{\overline{0}}$ is a marker automorphism of Σ_3.

EXAMPLE 5. The map ϕ_1 defined on Σ_3 by "interchange 1 and 2" whenever these digits are followed by a 1 is **not** an automorphism of Σ_3 since

$$\phi_1(111\ldots) = (222\ldots) = \phi(222\ldots),$$

so ϕ_1 is not one-to-one.

EXAMPLE 6. There are many different strings of symbols that can serve as markers and thereby generate marker automorphisms. For example, the string 220 generates a marker automorphism ϕ_{220} which interchanges 1 and 2 whenever these digits are followed by 220. For example

$$\phi_{220}(122022021220\ldots) = (222022022220\ldots)$$

EXAMPLE 7. We leave it to the reader to check that the strings 100 and $10\overline{0}$ both generate marker automorphisms in Aut_3. However, 101 does not generate an element of Aut_3. Of course, we can generate other marker automorphisms by interchanging the symbols 0 and 2, or the symbols 0 and 1, when they are followed by an appropriate marker.

How can we tell if a given string can serve to generate a marker automorphism? The general principle is that markers cannot "interfere" with previous markers in a string. We will make this more precise below when we give an algorithm to generate all markers for Aut_3.

3. Dynamics of Quadratic Polynomials

We now shift (pardon the pun) gears. In this section we recall some of the elementary properties of complex quadratic polynomials, including a description of their Julia sets and the Mandelbrot set. For more details on this material we refer to [**B1**], [**DK**], [**Mi**].

Let $P_c(z) = z^2 + c$ with $z, c \in \mathbb{C}$. It suffices to deal with quadratic maps in this special form as any quadratic polynomial is conjugate to some P_c. That is, if Q is any quadratic polynomial, we may find an affine map h of $\mathbb{C}$ and a particular c value for which we have $h \circ Q(z) = (h(z))^2 + c$. It follows easily that Q and P_c have the same dynamical structure as h carries Q-orbits to P_c-orbits. For example, the well known "logistic" quadratic polynomial $Q(z) = 4z(1-z)$ is conjugate to $P_{-2}(z) = z^2 - 2$ via the map $h(z) = -4z + 2$, as the reader may easily verify.

DEFINITION 1. The Julia set of P_c, denoted by J_c, is the boundary of the set of points whose orbit escapes to ∞.

Remark. J_c has several other equivalent definitions. For example, J_c is also the closure of the set of repelling periodic points as well as the set of points at which the family of iterates of P_c, $\{P_c^n\}$, fails to be a normal family. As a consequence of this final fact, if $z \in J_c$ and U is any neighborhood of z, then $\cup_n P_c^n(U)$ fills all of $\mathbb{C}$ missing at most one point. This is the famous theorem of Montel.

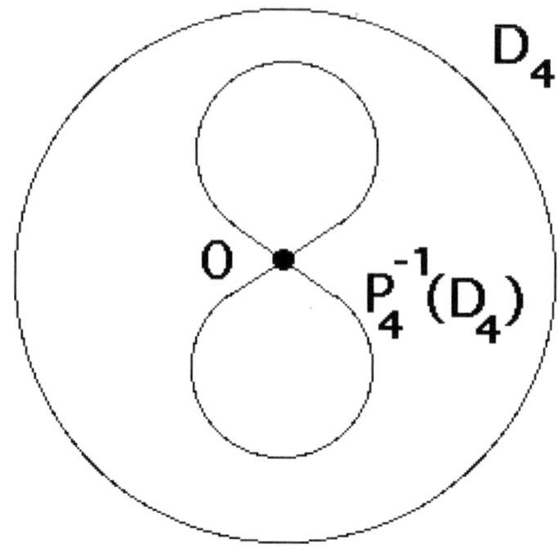

FIGURE 1. The case $c = 4$.

EXAMPLE 8. $P_0(z) = z^2$ has Julia set equal to the unit circle. Indeed, $P_c^n(z) \to \infty$ if and only if $|z| > 1$.

EXAMPLE 9. This second example will be crucial to what follows. Let $c = 4$. We claim that J_4 is a Cantor set. To see this, consider the closed disk D_4 of radius 4 centered at the origin. If $|z| \geq 4$, then $P_4^n(z) \to \infty$. This follows since

$$\begin{aligned}|z^2 + 4| &\geq |z|^2 - 4 \\ &\geq |z|^2 - |z| \\ &\geq (|z| - 1)|z| \\ &\geq 3|z|\end{aligned}$$

Hence J_4 is contained in the interior of D_4. To obtain J_4, we pull this disk back via preimages of P_4.

We claim that $P_4^{-1}(D_4)$ is a figure 8 region contained in the interior of D_4. Indeed, to obtain this preimage, we first subtract 4, obtaining a disk of radius 4 centered at -4. Then we take the square root, obtaining the result shown in Fig. 1. In particular, note that the maximum magnitude of any point in $P_4^{-1}(D_4)$ is $2\sqrt{2}$. Thus the preimage of the open disk $|z| < 4$ is contained in the two interior lobes of this figure eight. Hence J_4 lies in these two lobes which we denote by V_0 and V_1.

Note also that J_4 misses the closed disk $D_{1/2}$ of radius $1/2$ centered at 0, for this disk is mapped to a region that lies completely outside $V_0 \cup V_1$. Indeed, the image is a disk about 4 of radius only $1/4$. Since $|P_4'(z)| > 1$ in the exterior of $D_{1/2}$ it follows that $|P_4'| > 1$ on J_4.

So consider the two simply connected regions, $I_0 = V_0 - D_{1/2}$ and $I_1 = V_1 - D_{1/2}$. We have that $J_4 \subset I_0 \cup I_1$ and that either branch of P_4^{-1} restricted to I_j is a contraction. Let Q_j denote the branch of P_4^{-1} on D_4 taking values in I_j. $Q_j(I_0 \cup I_1)$ is a pair of small simply connected regions in I_j.

Now let $(s_0 s_1 s_2 \ldots) \in \Sigma_2$ and consider the subsets

$$I_{s_0 \ldots s_n} = Q_{s_0} \circ \ldots \circ Q_{s_{n-1}}(I_{s_n})$$

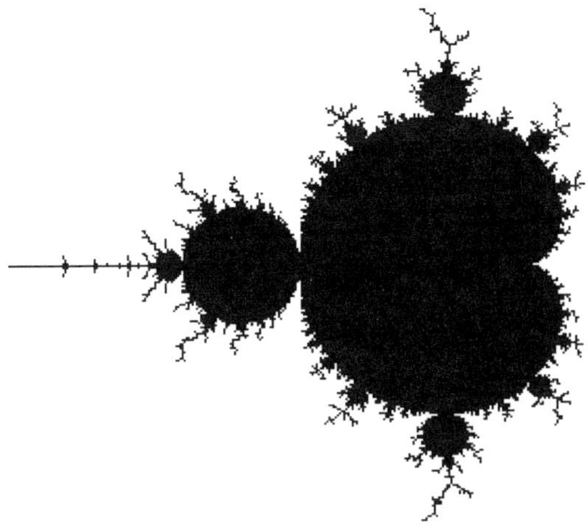

FIGURE 2. The Mandelbrot set.

One checks easily that

(1) $I_{s_0\ldots s_{n+1}} \subset I_{s_0\ldots s_n}$.
(2) $h(s_0s_1s_2\ldots) = \bigcap_{n=0}^{\infty} I_{s_0\ldots s_n}$ is a point (this uses the fact that the Q_j are contractions).
(3) h is a homeomorphism of Σ_2 onto J_4.
(4) $h(s_0s_1s_2\ldots) \in J_4$ has itinerary $(s_0s_1\ldots)$ relative to I_0 and I_1.

Remark. The crucial ingredient in example 9 is the fact that the orbit of the critical point tends to ∞. It can be shown that, whenever this occurs, J_c is a Cantor set. See [**B1**], [**Mi**]. On the other hand, if the orbit of 0 does not escape to ∞, then J_c is a connected set. Roughly speaking, this occurs since closed curves outside the Julia set never contain the critical value, and therefore their preimage is never a figure eight. Thus we have the *fundamental dichotomy* of quadratic dynamics: either J_c is a Cantor set, in which case $P_c^n(0) \to \infty$, or J_c is connected, in which case $P_c^n(c)$ is bounded. It is the well-known Mandelbrot set M (see Fig. 2) that gives a picture of this dichotomy: the parameter value c lies in M if and only if J_c is connected.

We now show how automorphisms of the shift arise naturally in complex dynamics. Given our extensive work with P_4 above, the construction is easy.

Consider $c = c(\theta) = 4e^{2\pi i\theta}$ so that c lies on the circle $|z| = 4$. For each such c, the corresponding Julia set is again a Cantor set, since $P_c^{-1}(D_4)$ is again a figure eight region well inside D_4. So we may invoke symbolic dynamics exactly as before. However, suppose we try to let I_0 and I_1 depend continuously on c. Since the I_j are essentially given via branches of the square root, it follows that, as θ moves completely around the circle, I_0 and I_1 only move in half-circles, i.e., they are interchanged. Thus, if we follow a path around this circle and consider a point with initial itinerary $s = (s_0s_1s_2\ldots)$, when we return to the initial c-value, all of the digits in s will have been changed. That is, we induce the only nontrivial automorphism in Aut_2 via this process. We call the map on the Julia set of P_4 induced by this process the *monodromy map.*

4. Polar Coordinates

We pause to introduce the first of the tools necessary for the construction of automorphisms, the Douady-Hubbard theory of external rays.

In a neighborhood of ∞, the value of c is more or less irrelevant and so P_c acts dynamically like z^2. More precisely, in a neighborhood of ∞, P_c is analytically conjugate to the squaring map $P_0(z) = z^2$. It is straightforward to check that the function

$$\phi_c(z) = \lim_{n\to\infty} (P_c^n(z))^{1/2^n}$$

is analytic and satisfies $\phi_c(P_c(z)) = (\phi_c(z))^2$ [**B1**], [**DK**]. Here we choose the principal branch of the square root. This enables us to view the dynamics of P_c near ∞ in "polar coordinates." Following Douady and Hubbard [**DH1**], we define the potential, or rate of escape function, by

$$h(z) = \lim_{n\to\infty} \frac{\log_+ |P_c^n(z)|}{2^n}.$$

The map h is continuous. Via the conjugacy, we see that

$$h(P_c(z)) = 2h(z).$$

Define

$$\Gamma_c = \{z \in \overline{\mathbb{C}} \,|\, h(z) > h(0)\}.$$

One checks easily that ϕ_c extends to a homeomorphism of Γ_c onto the complement of D_R where $R = \exp(h(0)) \geq 1$. If $r > R$, ϕ_c maps the level curve

$$h(z) = \log(r)$$

to the circle $|z| = r$. This gives us the radial component of our polar coordinate system.

For fixed t, the preimage under ϕ_c of the ray $\rho \mapsto \rho \exp(2\pi i t)$, $\rho > R$, is called the external ray of argument t for P_c. We denote this external ray by θ_t. The external rays give the angular coordinate on Γ_c.

We call the value $\log |\phi_c(c)|$ the *escape rate* of the critical value and $\operatorname{Arg} \phi_c(c)$ its external angle.

Note that, if the orbit of 0 escapes under P_c, then we cannot extend the conjugacy ϕ_c further. Indeed, each point in Γ_c has two preimages under P_c with one exception, namely the critical value c. Thus the level curve $h(z) = h(0)$ is a figure 8 curve in essentially the same manner as in the previous section. See Fig. 3

On the other hand, if the orbit of 0 does not escape to ∞, then $h(0) = 0$. Hence we can define ϕ_c everywhere in the exterior of the Julia set. This is the basic idea behind the proof that the Julia set is connected when the orbit of 0 remains bounded. More detail on all of this construction may be found in [**DH1**] or [**B1**], or the AMS Short Course lecture notes [**DK**], [**D**].

Note that the critical value c always lies in Γ_c. Hence $\phi_c(c)$ is always defined. This value plays a special role in complex dynamics. For one thing, $\phi_c(c)$ determines the quadratic polynomial up to affine conjugacy. That is, if two polynomials of the form $z^2 + c$ have the same escape rate and external angle, then these maps are identical. In addition, the mapping $\Phi(c) = \phi_c(c)$ is the external Riemann map that

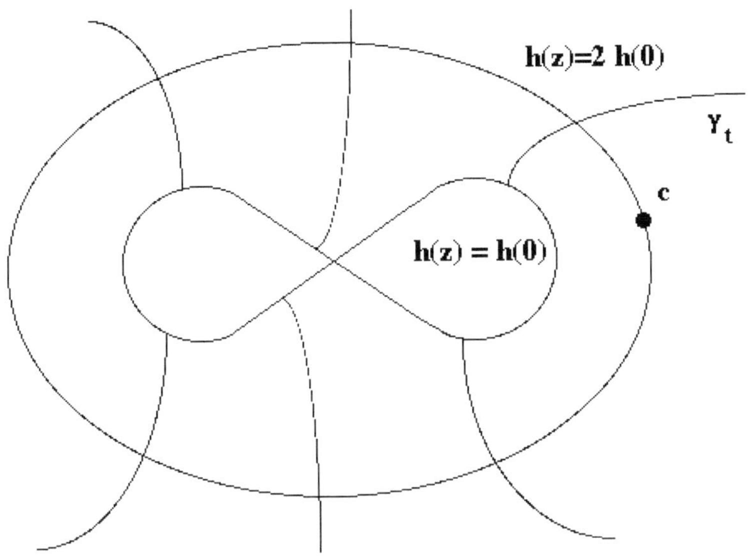

FIGURE 3. Polar coordinates.

takes the complement of the Mandelbrot set onto the exterior of the unit disk in $\mathbb{C}$. For more details, see [**DH1**].

5. The Measurable Riemann Mapping Theorem

Now we turn to our second tool, the Measurable Riemann Mapping Theorem (MRMT) of Ahlfors-Bers. This is a difficult and highly technical subject. Our treatment will be somewhat sketchy and without rigor, and will involve only the ideas we need to generate automorphisms of the shift. The idea is to convey just the flavor and power of this important result. Much more detail can be found in [**B1**], [**McM**].

Although this theory is quite general, we will restrict here to maps of the extended plane that are differentiable almost everywhere.

DEFINITION 2. An *ellipse field*or *complex structure* μ on $\mathbb{C}$ is a family of concentric ellipses that fills the tangent plane at each $z \in \mathbb{C}$.

Geometrically, we think of $\mu(z)$ as being a single (non-trivial) ellipse in the tangent plane at z, but technically $\mu(z)$ is the entire family given by multiplying the single ellipse by $t \in \mathbb{R}^+$. In our case, the ellipse fields will always vary smoothly a.e.

EXAMPLE 10. The *standard structure* μ_* on $\mathbb{C}$ is the field for which every ellipse is an actual circle. Note that a holomorphic map preserves μ_* in the sense that the derivative of this map sends circles in the tangent plane to circles in the image tangent plane. Indeed, a map is holomorphic if and only if it preserves μ_*.

EXAMPLE 11. A t-horizontal ellipse field is given by an ellipse at each z whose major axis lies parallel to the x-axis, whose minor axis lies parallel to the y-axis, and for which the ratio of the length of the major axis to the minor axis is $t > 1$.

EXAMPLE 12. Consider an annulus of the form

$$A = \{z \in \mathbb{C} \mid 0 < r_1 \le |z| \le r_2\}$$

with $r_1 < r_2$. The *Dehn twist map* T on A is given by

$$T(re^{2\pi it}) = r\exp(2\pi i(t + \frac{r - r_1}{r_2 - r_1})).$$

Note that T is the identity on the boundary of A and that T rotates each circle $|z| = r$ with $r_1 < r < r_2$ in the counterclockwise direction by an angle that increases from 0 to 2π as r increases from r_1 to r_2. We define a *twisted* complex structure on $\mathbb{C}$ to be the standard structure outside A and the image of μ_* under the derivative DT on A.

DEFINITION 3. The *dilatation* $d_\mu(z)$ of an ellipse field is the ratio of the length of the major axis to the minor axis.

So $d_\mu(z) \ge 1$ and $d_\mu(z) \equiv 1$ if and only if $\mu = \mu_*$. For the complex structures we consider, d_μ will always be bounded on $\mathbb{C}$.

DEFINITION 4. A homeomorphism $F \colon \mathbb{C} \to \mathbb{C}$ that is smooth almost everywhere is said to *straighten* an ellipse field μ if $DF(\mu) = \mu_*$ almost everywhere.

That is, F straightens μ if the map takes the given complex structure to the standard structure a.e. (It always seems strange that you "straighten" something by making it into a circle, but such is life in conformal geometry.) In the t-horizontal ellipse field above, we can straighten μ by applying the map $F(x, y) = (x/t, y)$. Similarly, we can straighten the twisted complex structure via a map that is the identity outside A and equal to the inverse of T on A. We do not worry about the structure on the boundary of A since we only require straightening a.e. We remark that such a straightening homeomorphism is called a *quasiconformal mapping* or a qc map for short.

In our framework, we can now state the celebrated Measurable Riemann Mapping Theorem of Ahlfors and Bers:

THEOREM 1. *(MRMT).* Suppose μ is an ellipse field on $\mathbb{C}$ with bounded dilatation. Then there is a homeomorphism F of $\mathbb{C}$ that straightens μ. Moreover, if we specify two values of F, then F is unique.

We mention that our restriction to smooth a.e. ellipse fields and straightening maps is a real restriction; the actual MRMT only requires that the ellipse field be measurable! It is amazing that you can straighten virtually anything with a qc map.

Here is the way that we will make use of the MRMT. Suppose $Q \colon \mathbb{C} \to \mathbb{C}$ is a map that preserves the complex structure μ. And suppose that F straightens μ. Then the mapping $P = F \circ Q \circ F^{-1}$ is a mapping of the complex plane that preserves the standard structure, by the chain rule. Thus P is a holomorphic map.

In particular, if we know that Q has finite degree, then P is a polynomial. In cases such as this, we say that Q is *quasiconformally conjugate* to the polynomial P.

6. Spinning the Critical Value

Now let's return to the quadratic example discussed above and use both the polar coordinates and the MRMT to generate the same result. The technique we will use is called *spinning the critical value* and will be the basic construction in the more difficult case of higher degree polynomials.

Throughout this section, we will fix $c \notin M$. We will construct a map that we will later show to be quasiconformally conjugate to P_c. Choose ρ_1, ρ_2 such that

$$h(c) < \rho_1 < \rho_2 < 2h(c)$$

and consider the annular region

$$A = \{z|\ \rho_1 \leq h(z) \leq \rho_2\}.$$

Let τ be the modified Dehn twist on A defined as follows. Let $r_i = e^{\rho_i}$ for $i = 1, 2$. Let

$$A' = \{z|\ r_1 \leq |z| \leq r_2\}.$$

Note that ϕ_c maps A onto A'. Define the usual Dehn twist T on the annulus A' as in the previous section. Then set

$$\tau(z) = \phi_c^{-1} \circ T \circ \phi_c(z).$$

Since the annular region A lies between the level sets of the potential containing c and $P_c(c)$, it follows that the inverse image $P_c^{-1}(A)$ is an annular region lying between the level sets of the potential containing 0 and c. Now define

$$F(z) = \begin{cases} P_c(z) & \text{if } z \notin P_c^{-1}(A) \\ \tau \circ P_c(z) & \text{if } z \in P_c^{-1}(A) \end{cases}$$

Clearly, F is a degree two branched cover which differs from P_c only on $P_c^{-1}(A)$. Although F is not a polynomial, the MRMT guarantees that it is quasiconformally conjugate to a polynomial — in fact, to P_c.

PROPOSITION 1. The map F is quasiconformally conjugate to P_c.

Proof. We define a new conformal structure μ on $\overline{\mathbb{C}}$ which is preserved by F as follows. If $h(z) \geq h(c)$, we set $\mu(z) = \mu_*(z)$. We then use F to pull back μ to $\overline{\mathbb{C}} - J_c$. That is, if $h(0) \leq h(z) < h(c)$, we define $\mu(z) = DF^{-1}(\mu(F(z))$. Note that there is no ambiguity here since the region $h(0) \leq h(z) < h(c)$ is disjoint from the region $h(z) \geq h(c)$. Note that $\mu = \mu_*$ everywhere in this region except in $P_c^{-1}(A)$. Continuing in this fashion we may define μ for all $z \in \overline{\mathbb{C}} - J_c$. By construction, F preserves μ. Since μ is the pullback of a complex structure with bounded dilatation by a map which is analytic except on $P_c^{-1}(A)$, it follows that μ has bounded dilatation on $\overline{\mathbb{C}} - J(P_c)$.

We extend μ to all of $\overline{\mathbb{C}}$ by setting $\mu = \mu_*$ on J_c. We may now apply the MRMT to obtain a quasiconformal homeomorphism f of $\overline{\mathbb{C}}$ which straightens μ a.e. Consequently, the map

$$Q = f \circ F \circ f^{-1}$$

preserves μ_* a.e. and so is analytic on $\overline{\mathbb{C}}$.

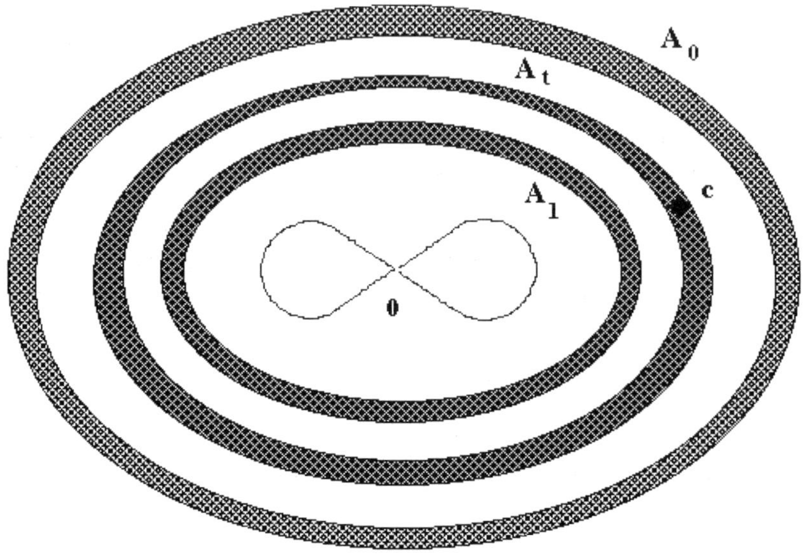

FIGURE 4. The annuli A_t.

We claim that Q is a quadratic polynomial which is affine conjugate to P_c. To see this, we may normalize f so that $f(\infty) = \infty$ and $f(0) = 0$. Thus, Q is a degree two map which fixes ∞ and which has branch points at 0 and ∞. Since Q preserves μ_*, Q is analytic. It follows that Q is a polynomial of degree two. Now $F = P_c$ on the region $h(z) > h(c)$, and f maps the critical orbit of F to that of Q. Hence the external angles and escape rates of the critical values of P_c and Q agree. Therefore, P_c and Q are equal. $\square$

We now use the same idea to construct a one-parameter family of maps F_t, $0 \leq t \leq 1$; we "spin the critical value around a level curve of the potential." We show that each member of this family is quasiconformally conjugate to a polynomial of the form $z^2 + c(t)$ where $c(t)$ is continuous and winds once around the Mandelbrot set for $0 \leq t \leq 1$. Hence this family induces the nontrivial automorphism of Σ_2 as in the special case $c = 4$ above.

Define a one-parameter family of annuli A_t for $0 \leq t \leq 1$ where $A_0 = \{z \mid \rho_1 \leq h(z) \leq \rho_2\}$ as above. For $0 \leq t \leq 1$, set $\eta_1(t) = (1 - \frac{1}{2}t)\rho_1$ and $\eta_0(t) = (1 - \frac{1}{2}t)\rho_2$. Let A_t denote the annular region

$$A_t = \{z | \ \eta_1(t) \leq h(z) \leq \eta_0(t)\}.$$

That is, A_t is the annular region whose inner and outer boundary curves are the h-level curves $\eta_1(t)$ and $\eta_0(t)$ respectively. Note that $A_0 = A$ and that A_1 is an annular region contained between the level sets $h(0) = \frac{1}{2}h(c)$ and $h(c)$. See Fig. 4. For each t we define a Dehn twist τ_t along the level curves of the potential in A_t exactly as we defined τ on A. Then we set

$$F_t(z) = \begin{cases} P_c(z) & \text{if } z \notin P_c^{-1}(A_t) \\ \tau_t \circ P_c(z) & \text{if } z \in P_c^{-1}(A_t) \end{cases}$$

Note that $F_0 = F$. Note also that there is an interval of t-values for which $c = P_c(0) \in A_t$. As t increases, it follows that the critical value $F_t(0) = \tau_t \circ P_c(0)$ is spun once around the level curve $h(z) = h(c)$. When $t = 1$, the critical value returns to its original location, i.e., $F_1(0) = c$.

PROPOSITION 2. *The map F_1 is also quasiconformally conjugate to P_c.*

Proof. The proof is the same as before: First define a new F_1-invariant complex structure as in the previous proposition. Use the MRMT with the same normalizations to straighten this structure via a quasiconformal homeomorphism f_1. Then $f_1 \circ F_1 \circ f_1^{-1}$ is a polynomial of degree two. Note that f_1 is analytic on $\{z \mid h(z) > h(c)\}$ and preserves the critical orbit. Hence $f_1 \circ F_1 \circ f_1^{-1}$ is quasiconformally conjugate to P_c as before. □

PROPOSITION 3. *There is a continuous function $c(t)$, $0 \leq t \leq 1$, such that:*

1. *F_t is quasiconformally conjugate to $z^2 + c(t)$ for each t;*
2. *$h(c(t))$ is constant; and*
3. *the external argument of $c(t)$ increases monotonically from 0 to 1 as t increases from 0 to 1.*

Proof. The crucial observation here is that the external rays for P_c are identical to those of F_t in the exterior of $P_c^{-1}(A_t)$. However, the location of the critical value $\tau_t \circ P_c(0)$ relative to these rays changes. Indeed, $\tau_t \circ P_c(0)$ passes through each ray exactly once as t increases. Thus we may invoke the preceding arguments to show that F_t is quasiconformally conjugate to $z^2 + c(t)$ where $c(t)$ satisfies 2 and 3. □

Consequently, the polynomials which are conjugate to F_t lie on a loop that winds once around M. As in the previous section, the monodromy map around this loop induces the non-trivial automorphism of Σ_2.

7. Higher Degree Polynomials

Let P be a polynomial of degree d. The Julia set, J_P, is defined exactly as in the quadratic case: J_P is the boundary of the set of points whose orbits tend to ∞. If the orbits of all critical points tend to ∞, then the Julia set of P is totally disconnected and $P|J_P$ is topologically conjugate to the d-shift. The proof of this is analogous to the proof in the quadratic case; see [**B1**], [**Mi**] for details.

The fundamental dichotomy for quadratic polynomials no longer holds when $d \geq 3$. Nevertheless, there is an intimate relationship between the connectivity properties of the Julia set and the orbits of the finite critical points. For example, if all of these orbits are bounded, then the Julia set is connected. If at least one critical point iterates to infinity, then the Julia set is disconnected, but it may or may not be totally disconnected.

For example, consider the situation when $d = 3$. A generic cubic has two distinct critical points. There are three cases:

1. Both critical orbits are bounded. Then J_P is connected.
2. Both critical points tend to ∞. Then J_P is a Cantor set, and the dynamics of the polynomial on J_P is conjugate to the 3-shift.
3. One critical point tends to ∞ and the other has a bounded orbit. When this happens, the Julia set is disconnected, and it may even be totally-disconnected.

Let X_d be the set of all monic, centered polynomials of degree d. That is, all polynomials of the form

$$z^d + a_{d-2}z^{d-2} + a_{d-3}z^{d-3} + \ldots + a_1 z + a_0$$

where the $a_j \in \mathbb{C}$. As in the quadratic case, any degree-d polynomial is conjugate to a polynomial of this form. Hence the parameter space for X_d has complex dimension $d-1$, the same as the number of critical points for a "typical" polynomial in X_d.

Let S_d be the escape locus, i.e., the subset of X_d consisting of polynomials whose critical points all escape to infinity. If $P \in S_d$, the Julia set of P is a Cantor set and P is conjugate to the d-shift on $J(P)$. S_d is open in X_d. We will think of the boundedness locus $X_d - S_d$ as the analogue of the Mandelbrot set for degree d polynomials, although one might argue that this analogue should really be the set of polynomials whose critical orbits are all bounded. Nonetheless, we will consider loops that lie in S_d and circle around various arms of $X_d - S_d$ that extend off to ∞.

To accomplish this, we generalize the spinning construction of the previous section. The ideas here are basically the same as in the quadratic case; the presence of additional critical points adds only some minor technicalities. As in the quadratic case, if $P \in X_d$, there is a neighborhood U of infinity and an analytic homeomorphism $\phi : U \to \overline{\mathbb{C}} - D_r$ such that $\phi \circ P(z) = (\phi(z))^d$. That is, any polynomial in X_d is conjugate near ∞ to the map $z \mapsto z^d$. As for quadratics, define the potential function

$$h_P(z) = \lim_{n\to\infty} \frac{\log_+ |P^n(z)|}{d^n}$$

For $z \in U$, let $h_P(z) = \log|\phi(z)|$. Let $c_1, \ldots, c_{d-1}$ be the finite critical points of P and $v_1, v_2, \ldots, v_{d-1}$ be the corresponding critical values. Then S_d can be characterized by

$$S_d = \{P \in X_d \,|\, h_P(c_i) = \frac{1}{d} h_P(v_i) > 0 \text{ for all } i = 1, \ldots, d-1\}.$$

We will always assume that c_1 is the critical point with the slowest escape rate, i.e., $h_P(c_1) \leq h_P(c_j)$ for $j = 2, \ldots, d-1$.

Let $P_i \in S_d$ for $i = 1, 2$, and define

$$U_i = \{z \,|\, h_i(z) \geq h_i(c_1)\}$$

where h_i, c_1 are the corresponding potentials and lowest critical points for P_i. Here we have suppressed the dependence of c_1 on i.

The next Proposition is proved more or less just as in the quadratic case. See [**BDK**] for more details.

PROPOSITION 4. *Let P_i, for $i = 1, 2$, be two polynomials in S_d. If $f : U_1 \to U_2$ is a quasiconformal conjugacy, then f can be extended to a quasiconformal conjugacy defined on all of $\overline{\mathbb{C}}$.*

Now we move to the spinning construction for higher degree polynomials. Let $P \in S_d$ and let h be the associated potential. Suppose that $h(c_1) < h(c_j)$ for $j = 2, \ldots, d-1$. Thus there exists $\epsilon > 0$ such that c_1 is the only critical point in the region $\{z \,|\, h(z) \leq h(c_1) + \epsilon\}$. We also assume that $d^n h(c_1) \neq h(c_j)$ for all $j > 1$ and n. These two assumptions serve to isolate the level sets of the potential corresponding to the orbit of c_1 from all of the other critical levels. Since these

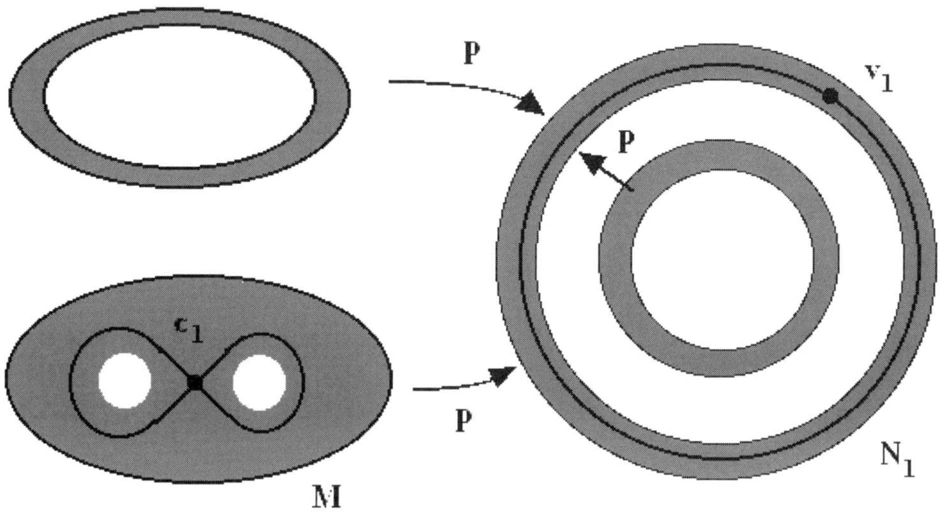

FIGURE 5. The components of $P^{-1}(N_1), M, N_1$.

level sets are dynamical invariants, this in turn allows us to guarantee the existence of quasiconformal conjugacies during the spinning construction.

Let γ denote the component of the level set $h^{-1}(h(v_1))$ which contains v_1. There are precisely $d-1$ components of $P^{-1}(\gamma)$. We denote them by $\alpha_1, \ldots, \alpha_{d-1}$ and assume that α_1 is the component containing c_1. Hence α_1 is a figure eight, while the remaining α_j are simple closed curves. We may choose $\epsilon > 0$ small enough so that the region

$$N = \{z | h(v_1) - \epsilon \leq h(z) \leq h(v_1) + \epsilon\}$$

is disjoint from all of the other critical level curves

$$\{z \,|\, h(z) = \frac{1}{d^m} h(v_k)\}, \; m = 1, 2, \ldots.$$

Let N_1 denote the component of N containing v_1 (and hence γ). It follows that N_1 is an annular region. Let M denote the component of $P^{-1}(N_1)$ that contains c_1. See Fig. 5.

In analogy with the construction in the quadratic case, we will modify P on M. For each $t \in [0, 1]$, let A_t be the annular region

$$A_t = \{z \in N_1 \,|\, h(v_1) - \epsilon t \leq h(z) \leq h(v_1) + \epsilon(1-t)\}$$

Note that A_1 is an annular region bounded by γ on the outside while A_0 is bounded by γ on the inside.

As in Section 6, there is a conformal map

$$\phi_t : A_t \to A_t' = \{\xi \,|\, 1 \leq |\xi| \leq r_0(t)\}$$

and we may choose ϕ_t so that these maps depend continuously on t. Let

$$T_t : A_t' \to A_t'$$

be a Dehn twist, i.e.,

$$T_t(re^{2\pi i s}) = r \exp(2\pi i(s + \frac{r-1}{r_0 - 1})).$$

Let $\tau_t = \phi_t^{-1} \circ T_t \circ \phi_t$. Now define

$$F_t(z) = \begin{cases} P(z) & z \notin M \cap P^{-1}(A_t) \\ \tau_t \circ P(z) & z \in M \cap P^{-1}(A_t) \end{cases}$$

PROPOSITION 5. *For all $t \in [0,1]$, there is a polynomial Q_t such that Q_t is quasiconformally conjugate to F_t.*

Proof. As above, we define a new conformal structure μ_t which is preserved by F_t. Let μ_* be the standard conformal structure. We set $\mu_t = \mu_*$ in the region

$$\{z \mid h(z) > h(v_1) - \epsilon(1-t)\}.$$

Let μ_t be the pullback of μ_* by F_t. This defines μ_t everywhere except on J_P. By definition, F_t preserves μ_t.

We define μ_t on J_P by setting $\mu_t = \mu_*$. Since μ_t is given by pulling back a complex structure with bounded dilatation by a polynomial, it follows that μ_t is a complex structure with bounded dilatation as well.

Now apply the MRMT. This gives a quasiconformal homeomorphism f_t which straightens μ_t. We may normalize so that $f_t(\infty) = \infty$. Then we have $f_t \circ Q_t \circ f_t^{-1} = Q_t$ is an analytic map of degree d which has a super attracting fixed point at ∞. Therefore, Q_t is a polynomial. If we further normalize so that $f_t'(\infty) = 1$ and $f_t(0) = 0$, then it follows that Q_t is monic and centered. □

PROPOSITION 6. *In the above construction, both $Q_0 = Q_1 = P$.*

Proof. We will prove this by writing down an explicit conjugacy g_i between F_i and P for $i = 0, 1$. These conjugacies will be quasiconformal. Moreover, they will vanish at 0 and be equal to the identity on a neighborhood of ∞. Hence they are the conjugacies that we obtained by the MRMT in the above argument.

We begin with F_0. Define $\Gamma_r = \{z \mid h(z) > r\}$. Define g_0 to be the identity map on $\Gamma_{h(c_1)+\epsilon}$. Clearly, we have $g_0 \circ F_0 = P \circ g_0$ on this region.

Let $B \subset M$ be the component of $P^{-1}(A_0)$ which contains c_1. The interior of B is an open annulus. There are two smooth maps $\zeta_i : B \to B$, $i = 1, 2$, that satisfy $P \circ \zeta_i = \tau_0 \circ P$. Each of the ζ_i fixes one of the boundary curves of B and rotates the other by a half twist. We choose τ_0' to be the ζ_i which fixes the outer boundary of B. Set $g_0 = \tau_0'$ on B. Also define g_0 to be the identity on $\Gamma_{h(c_1)} - B$. It follows that $P \circ g_0 = g_0 \circ F_0$. We now extend g_0 to lower h-levels in the natural way. If z satisfies

$$\frac{1}{d} h(c_1) \leq h(z) < h(c_1)$$

we set $g_0(z) = P^{-1} \circ g_0 \circ F_0(z)$ where we choose the appropriate branch of the inverse of P^{-1} to make g_0 continuous.

It is important to note that $P^{-1} \circ F_0$ is not the identity on the two components of the interior of the figure eight component of α_1. Indeed, g_0 interchanges these two components while preserving the interiors of all other components.

Now continue as in the quadratic case. This defines a quasiconformal homeomorphism of $\mathbb{C} - J_P$ which extends quasiconformally to all of $\overline{\mathbb{C}}$ as in Proposition 4. Hence P is quasiconformally conjugate to F_0 via a conjugacy that is the identity on a neighborhood of ∞. Furthermore, g_0 fixes the critical points of P. Hence $P = Q_0$.

We now turn to Q_1. We remark that Q_1 is affine conjugate to P by Proposition 5 since f_1 is conformal on $\Gamma_{h(c_1)}$. We prefer, for later purposes, to construct f_1 directly. To do this, define g_1 = identity on $\Gamma_{h(c_1)}$. Let B be the component of $P^{-1}(A_1)$ which contains c_1. Note that, unlike the previous case, the interior of B consists of two disjoint annuli each mapped isomorphically onto B by P. We can choose a map τ_1' on each of these components so that $P \circ \tau_1' = \tau_1 \circ P$ since P is an isomorphism on each component. Hence we set $g_1 = \tau_1'$ on B. Define g_1 to be the identity on $\Gamma_{h(c_1)-\epsilon} - B$ and then continue as before. It follows again that g_1 is a quasiconformal homeomorphism conjugating P and Q_1. Note that, unlike the previous case, g_1 preserves all components of $\mathbb{C} - \Gamma_{h(c_1)-\epsilon}$. $\square$

COROLLARY 1. *As t decreases from 1 to 0, the components of the interior of $h_{Q_t}^{-1}(h(c_1))$ containing c_1 are interchanged.*

8. The Cubic Case: Four Examples

Rather than give the general construction of automorphisms in the cubic case, we will restrict our attention to a number of special but nonetheless illustrative examples. All are contained in the one-parameter family

$$P_a(z) = z^3 - 3a^2 z + 5$$

where $a \in \mathbb{R}$. The critical points of P_a are $\pm a$ and the value of the constant term is of little significance other than its being large enough to display all of the desired cases. The five examples that we discuss are defined by the graphs in Fig. 6. In particular, the a-values are determined by these graphs. Let $h_a(z)$ denote the corresponding potential function.

We show how to generate specific automorphisms of the 3-shift by applying the spinning construction to each of these examples.

Case 1: $a = 0$. Note that $P_0(z) = z^3 + 5$ has a unique critical point at 0. The level sets for the potential for P_0 containing the critical point and value are shown in Fig. 7. As in the quadratic case, the critical value lies on a simple closed curve. The preimage of this curve is mapped in three-to-one fashion onto this curve, except, of course, at 0, which is the only preimage of 5. Hence the preimage is a curve which has three lobes pinched at 0. We denote the 3 components of $h_0(z) < h_0(0)$ by 0, 1 and 2 respectively. This is analogous to the quadratic case, where we labeled the two components I_0 and I_1. Note that there is an arbitrary choice involved in this selection. If we now use the spinning construction of the previous section to spin the critical value around the level set $\gamma = h_0^{-1}(h_0(5))$, then we induce a one-third turn on the lobes of $\alpha = h_0^{-1}(h_0(0))$, and this yields an automorphism that cyclically permutes the symbols. This, of course, follows just as in the quadratic case. Now, however, by changing a, we can separate the critical points and thereby change the configuration of the level sets.

Case 2: $a = a_1$. According to Fig. 6, the orbit of c_1 escapes a little more more slowly than the orbit of c_2 when $a = a_1$. More precisely, we see that

$$h_a(c_1) < h_a(c_2) < h_a(v_1).$$

The corresponding critical levels of the potential are shown in Fig. 8. The critical values now lie on distinct level sets, each of which is a simple closed curve. Preimages of level sets for which $h_a(z) > h_a(v_2)$ are simple closed curves. When the value

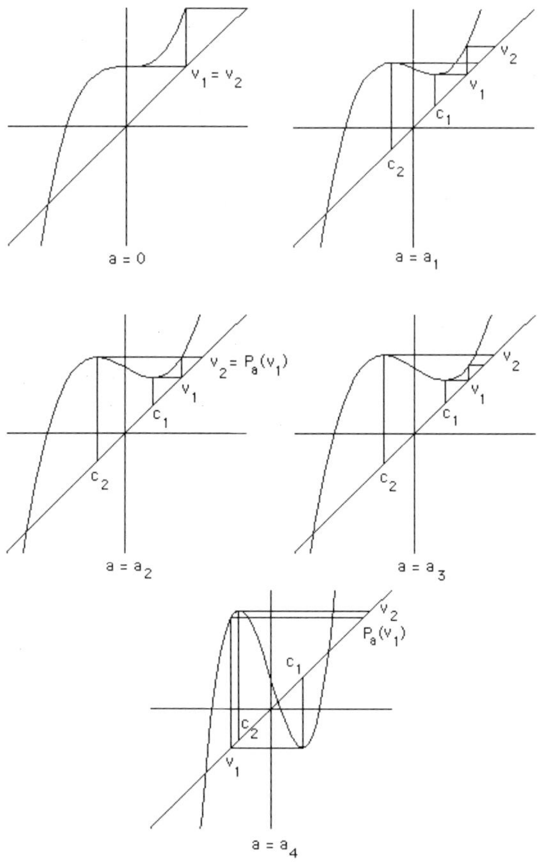

FIGURE 6. Graphs of the P_a.

of the potential reaches $h_a(v_2)$, the preimage becomes pinched at the critical point c_2. One of the lobes inside this level curve is mapped in two-to-one fashion onto $h_a(z) \leq h_a(v_2)$. This corresponds to a region containing two of the symbols from the previous case, say 1 and 2. Thus we denote this curve by $\overline{0}$. This is the lobe containing c_1. The other lobe is mapped in one-to-one fashion onto $h_a(z) \leq h_a(v_2)$; we denote this lobe by 0.

Now the preimage of the level curve containing v_1 consists of two distinct curves. One is a figure eight curve that contains c_1 and lies in $\overline{0}$; the other is a simple closed curve in 0. The region $\{z \,|\, h_a(z) < h_a(c_1)\}$ may again be used to describe the symbolic dynamics on the Julia set. In analogy with the quadratic case, this labeling indicates the existence of a homeomorphism between the Julia set J and Σ_3 where each point in J contained in the disk labeled 0 (or 1 or 2 respectively) is labeled with a sequence starting with the symbol 0 (or 1 or 2 respectively).

If we now spin the critical value v_1 around the curve $\gamma = h_a^{-1}(h_a(v_1))$, the resulting automorphism simply interchanges every 1 and 2; 0 is left fixed.

Case 3: $a = a_2$**.** This is the special case where $v_2 = P_a(v_1)$ and

$$h_a(c_1) < h_a(c_2) = h_a(v_1) < h_a(v_2).$$

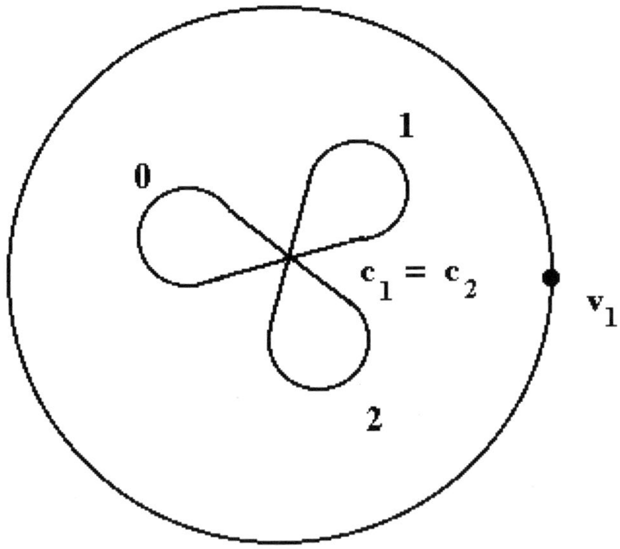

FIGURE 7. Level curves for $a = 0$.

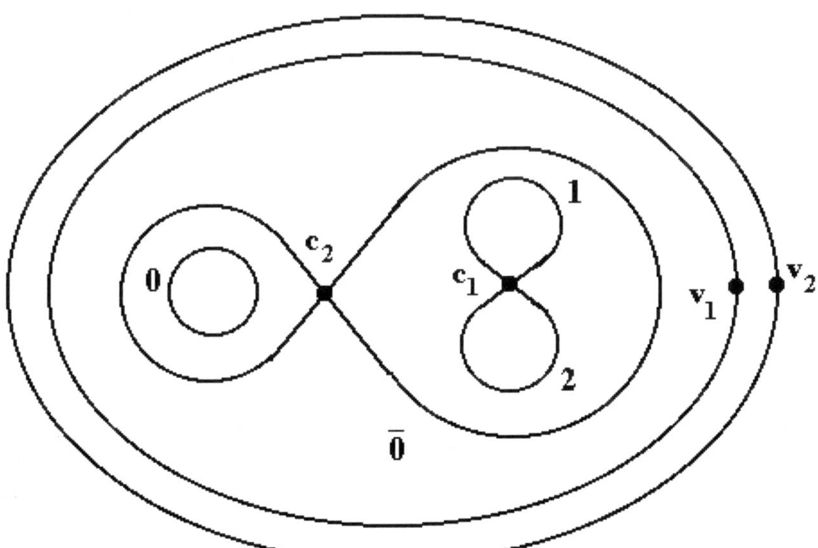

FIGURE 8. Level curves for $a = a_1$.

As above, the level set $h_a(z) = h_a(c_2)$ is a figure eight curve with one lobe (previously called $\overline{0}$) mapped in two-to-one fashion onto the image, and the other mapped in one-to-one fashion. By assumption, v_1 lies on this curve, but not at c_2. Let us assume that v_1 lies on the boundary of the lobe containing c_1. (Actually, given the graph of P_a, this must be the case.) This forces some additional pinching in the

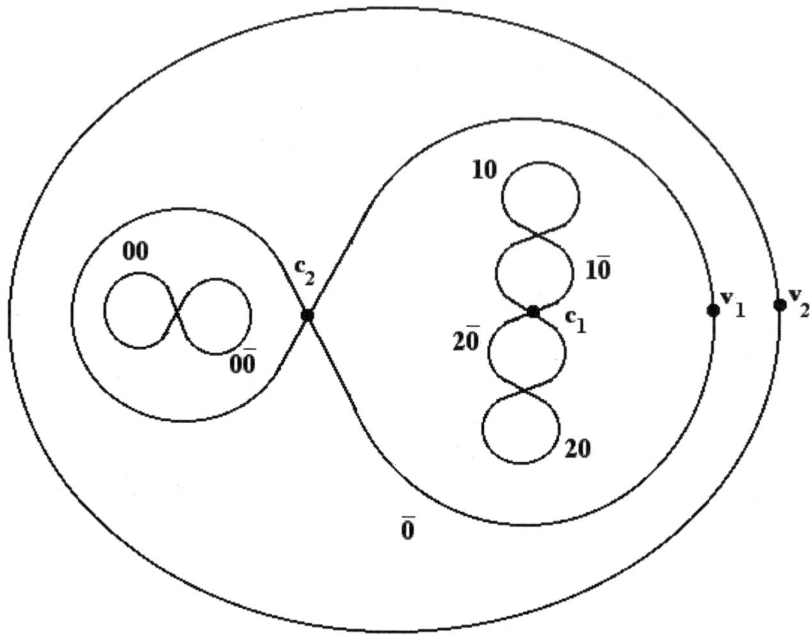

FIGURE 9. Level curves for $a = a_2$.

level set $h_a(z) = h_a(c_1)$; we get pinching at c_1 as well as at the two preimages of c_2. That is, the region $h_a(z) \leq h_a(c_1)$ now consists of four lobes in $\overline{0}$. Two of these lobes are mapped to the lobe called 0; these we denote by 10 and 20. The other two lobes map onto $\overline{0}$; these we denote by $1\overline{0}$ and $2\overline{0}$. These level curves are depicted in Fig. 9. In the region 0, the preimage of $h_a(z) = h_a(c_2)$ is a figure eight curve which is mapped in one-to-one fashion onto $h_a(z) = h_a(c_2)$. Thus one of these lobes is denoted 00, while the other (the lobe mapped onto $\overline{0}$) is denoted $0\overline{0}$.

We will not spin the critical value v_1 at this level as the critical value does not lie on a simple closed level curve.

Case 4: $a = a_3$. Now suppose we change the parameter a so that

$$h_a(v_1) < h_a(c_2) < h_a(P_a(v_1)) < h_a(v_2).$$

From the graph, it follows that v_1 now lies slightly inside the level curve containing c_2. The level set $h_a(z) = h_a(v_1)$ now consists of two simple closed curves, one inside each lobe of $h_a(z) = h_a(c_2)$. Then the multiply pinched critical level curves of $h_a(z)$ then break apart as shown in Fig. 10. Inside $\overline{0}$, we have a figure eight curve containing c_1 and bounding the regions $1\overline{0}$ and $2\overline{0}$. This region is mapped onto the level set containing v_1. There are two other preimages inside $\overline{0}$, namely 10 and 20; these are mapped onto the other component of $h_a(z) = h_a(v_1)$, specifically the one in the region 0.

If we now spin the critical value v_1 around its level curve, then only the components of the Julia set marked $1\overline{0}$ and $2\overline{0}$ are interchanged; the components 10 and 20 are left fixed. Thus, spinning induces the automorphism given by the marker set $\overline{0}$: interchange 1 and 2 whenever followed by 1 or 2.

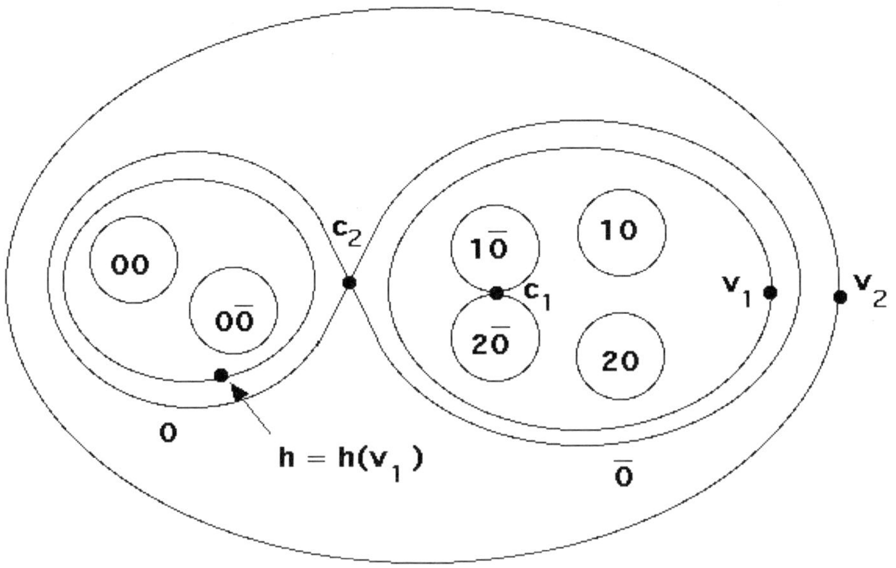

FIGURE 10. Level curves for $a = a_3$.

At this point we note that we have reached the polynomial P_{a_4} by simply "pushing" the critical value v_1 down the real axis. Note that $h_a(v_1)$ decreases during this process. This will be the essential ingredient in the next section when we show how to perform this "pushing deformation" in general.

9. The pushing deformation.

In passing from $a = a_1$ to $a = a_3$ in the previous examples, we pushed the critical value v_1 down the real axis across the level set $h_a(z) = h_a(c_2)$. This is the second construction involving the MRMT. In the general case we will push the critical value v_1 through the level set $h(z) = 3^{-k}h(c_2)$ for each $k \geq 0$.

To be precise, let r satisfy $h_a(c_2) < h_a(v_1) < 3r < h_a(v_2) < 9r$, or, equivalently, $h_a(c_1) < r < h_a(c_2) < h_a(v_1) < 3r$. Consider the region

$$S = \{z \,|\, r \leq h_a(z) \leq 3r\}.$$

Note that both v_1 and c_2 lie in S. See Fig. 11. We can construct a homeomorphism f of S that takes v_1 to a point that lies inside one of the components defined by the $\overline{0}$ lobe of the level curve $h_a(z) = h_a(c_2)$. Let S' denote the preimage of S that contains c_1.

Then define

$$F(z) = \begin{cases} P_a(z) & \text{if } z \in \overline{\mathbb{C}} - S' \\ f \circ P_a(z) & \text{if } z \in S'. \end{cases}$$

Note that $F(z)$ is conformal outside of S'. Next we define (in the usual manner) an F-invariant quasiconformal structure on $\overline{\mathbb{C}} - S'$. We pull this structure back to $\overline{\mathbb{C}} - J_{P_a}$ using F. Finally, we extend the structure to all of $\overline{\mathbb{C}}$ by setting it equal to the standard structure on J_{P_a}. This structure has bounded dilatation, since F

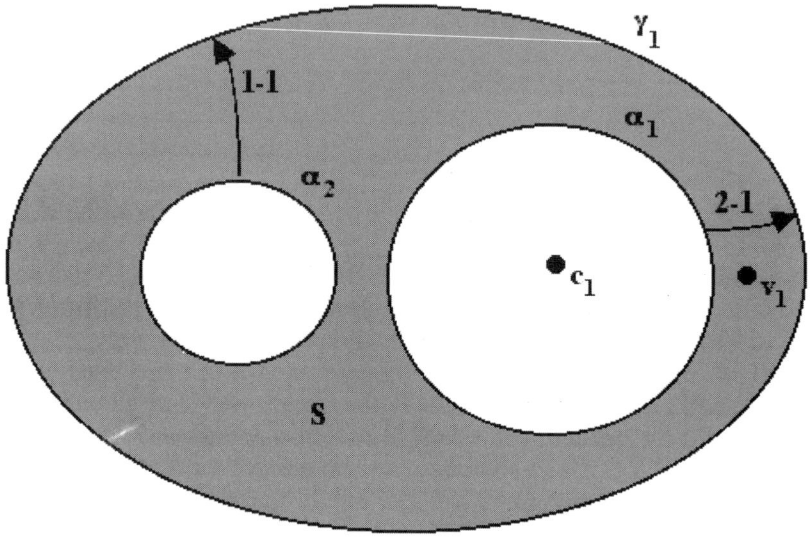

FIGURE 11. The region S.

is conformal except on S' and the pullback from S to S' introduces only a finite amount of distortion. By applying the MRMT to this new F-invariant structure, we obtain a quasiconformal homeomorphism $f : \overline{\mathbb{C}} \to \overline{\mathbb{C}}$ such that $f^{-1} \circ F \circ f$ is a polynomial P_f. Now, given P_f, we can perform the spinning construction as above to make v_1 spin as in Case 4.

It is important to note that the above construction is analytic in its parameters (see [AB]). For example, if we apply it to a one-parameter family f_t which varies analytically in t, then the resulting polynomials P_{f_t} also vary analytically in t. In essence, the pushing construction allows us to deform the polynomial continuously using a quasiconformal deformation of the Riemann surface S.

Now, using the pushing deformation, we could have equally well moved v_1 into the other component defined by $h(z) = h(v_1)$. This brings us to our last example.
Case 5: $a = a_4$. The levels for this a-value satisfy

$$h_a(v_1) < h_a(c_2) < h_a(P_a(v_1)) < h_a(v_2)$$

as before, but note that the critical value v_1 is now in the other component of $h(z) = h(v_1)$. The level sets for h_a inherit a different marking in this case. See Fig. 12.

Spinning the critical value here interchanges the components marked 10 and 20, i.e., it induces the automorphism with marker 0.

Exercise. In cases 4 and 5 above, continue pushing the critical value deeper and identify the corresponding automorphisms that are generated. Specifically, first consider the case where $h_a(P_a(v_1)) = h_a(c_2)$ and sketch the level sets of the potential. Then investigate what happens when the critical value descends further.

If we continue inductively by pushing and spinning, we create a tree whose edges correspond to the pushing deformation and whose vertices correspond to polynomials in the escape locus from which we begin the spinning process. This

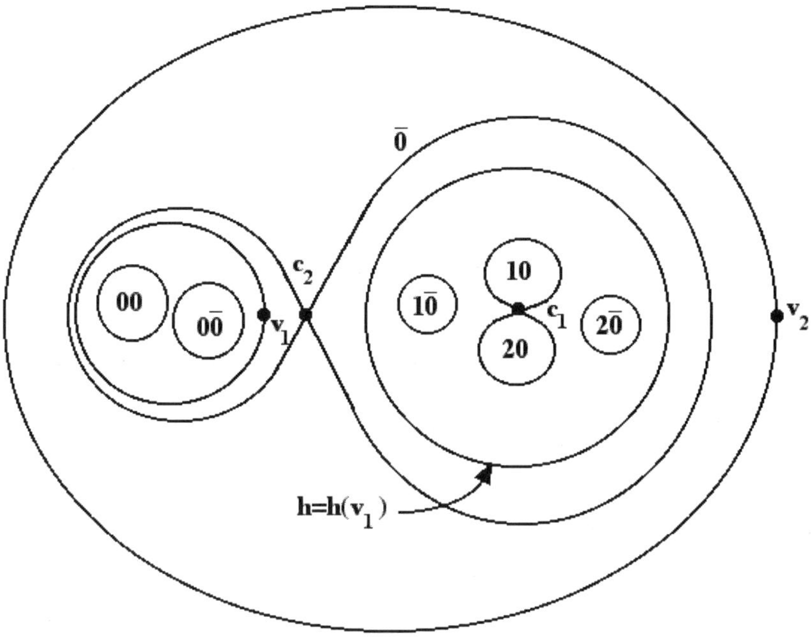

FIGURE 12. Level curves for $a = a_4$.

tree thus generates a collection of automorphisms. Using results of Ashley [**Ash**], we have proved in [**BDK**]:

THEOREM 2. The tree generated by following all possible topologically distinct paths in the cubic escape locus yields a complete set of generators for Aut_3. A similar statement holds for Aut_d.

Table 1 lists the minimal markers and marker sets of length j for $j = 1, 2, 3, 4$. These correspond to the vertices of T at depth j

Length 1:	0	$\overline{0}$				
Length 2:	00	$0\overline{0}$	10	20	$\overline{00}$	
Length 3:	000	$00\overline{0}$	$0\overline{0}0$	$01\overline{0}$	$02\overline{0}$	100
	$10\overline{0}$	200	$20\overline{0}$	110	120	210
	220	$\overline{000}$				
Length 4:	0000	$000\overline{0}$	$00\overline{0}0$	$002\overline{0}$	$001\overline{0}$	0100
	0200	$0\overline{0}0\overline{0}$	$01\overline{0}0$	$011\overline{0}$	$012\overline{0}$	$02\overline{0}0$
	$021\overline{0}$	$022\overline{0}$	1000	$100\overline{0}$	1010	1020
	$10\overline{00}$	2000	$200\overline{0}$	2010	2020	$20\overline{00}$
	1100	$110\overline{0}$	1200	$120\overline{0}$	2100	$210\overline{0}$
	2200	$220\overline{0}$	1110	1210	2110	2210
	1120	2120	1220	2220	$\overline{0000}$	

Table 1

10. Conclusion

The natural question that arises involves the two-sided d-shift map. In this case the sequence space is the set of two sided sequences $(\dots s_{-2}s_{-1}.s_0s_1s_2\dots)$ where $0 \leq s_j \leq d-1$, and the shift map σ really is a shift in this case, as σ simply shifts the decimal point one spot to the right:

$$\sigma(\dots s_{-2}s_{-1}.s_0s_1s_2\dots) = (\dots s_{-2}s_{-1}s_0.s_1s_2\dots)$$

The question here is: What is the structure of the group of automorphisms of this shift? It is known that, even in the case $d = 2$, this is a very large group.

We conjecture that the answer to this question is intimately related (in the case $d = 2$) to the family of maps known as the Hénon map. This family of maps is given by

$$\begin{aligned} x_1 &= A - By_0 - x^2 \\ y_1 &= x. \end{aligned}$$

Note that the Jacobian of this map is identically equal to B; this is what makes this map special. In the plane, it is known that, provided A is large enough, this map has an invariant set homeomorphic to a Cantor set on which the dynamics are equivalent to the two-sided shift map. See [**DN**]. Indeed, this is a simple nonlinear version of the classic Smale horseshoe map. If we move to $\mathbb{C}^2$ (making the two parameters A and B as well as x_0 and y_0 complex, then Hubbard and Oberste-Vorth [**HOV**] have shown that similar results apply. Moreover, we can now consider the set of parameters for which we have such a Cantor set (the analogue of the escape locus, although there are no critical points around). As before, moving around loops in this locus also induces non-trivial automorphisms of the two-sided shift. The big question then is: Can we generate all automorphisms of the two-sided shift in this way?

References

[A] Ahlfors, L. *Lectures on Quasiconformal Mappings,* Van Nostrand, 1966.

[AB] Ahlfors, L. and Bers, L. The Riemann Mappings Theorem for Variable Metrics, *Annals of Math.*, Vol. 72-2, 1960, 385-404.

[Ash] Ashley, J. Marker Automorphisms of the One-Sided Shift, *Ergodic Theory and Dynamical Systems*, Vol. 10, 247-262.

[BDK] Blanchard, P., Devaney, R. L., and Keen, L. The Dynamics of Complex Polynomials and Automorphisms of the Shift. *Invent. Math.* **104** (1991), 545-580.

[BFK] Boyle, M., Franks, J., and Kitchens, B. Automorphisms of the One-Sided Shift and Subshifts of Finite Type. *Ergodic Theory Dynam. Systems.* **10** (1990), 421-449.

[B1] Blanchard, P. Complex Analytic Dynamics on the Riemann Sphere, *B.A.M.S.* Vol. II, No.1, 1984, 85-141.

[B2] Blanchard, P. Disconnected Julia Sets, *Chaotic Dynamics and Fractals*, Ed. M. Barnsley, S. Demko, Academic Press, (1986), pp. 181-201.

[BH1] Branner, B. and Hubbard, J. The Iteration of Cubic Polynomials I: The Global Topology of Parameter Space. *Acta. Math.* **160** (1988), 143-206.

[BH2] Branner, B. and Hubbard, J. The Iteration of Cubic Polynomials II: Patterns and Parapatterns. *Acta. Math.* **169** (1992), 229-325.

[D] Devaney, R. L., ed. *Complex Dynamical Systems: The Mathematics Behind the Mandelbrot and Julia Sets.* American Mathematics Society, 1994.

[DK] Devaney, R. L. and Keen, L.. eds. *Chaos and Fractals: The Mathematics Behind the Computer Graphics.* American Mathematics Society, 1989.

[DH1] Douady, A. and Hubbard, J. Itération des Polynômes quadratiques complexes, *C.R. Acad. Sci. Paris*, t.29, Serie I-1982, pp. 123-126.

[DH2] Douady, A. and Hubbard, J. Étude Dynamique des Polynôme Complexes, *Publications Mathematiques d'Orsay*, v. 84-102.

[DH3] Douady, A. and Hubbard, J. On the Dynamics of Polynomial-like Mappings, *Ann. Scient., Éc. Norm. Sup.* 4e *séries*, t.18, (1985), 287.

[DN] Devaney, R. L. and Nitecki, Z. Shift Automorphisms in the Hénon Mapping. *Comm. Math. Phys.* **67** (1979), 137-146.

[F] Fatou, P. Sur les Equations Fonctionnelles, *Bull. Soc. Math. France* **47** (1919), 161-271.

[Fr] Franks, J. *Homology and Dynamical Systems.* CBMS Regional Conference Series. **49** (1982).

[GK] Goldberg, L. and Keen, L. The Mapping Class Group of a Generic Quadratic Rational Map and Automorphisms of the Two Shift. *Invent. Math.* **101** (1990), 335-372.

[H] Hedlund, G. Endomorphisms and Automorphisms of the Shift Dynamical System. *Math. Syst. Theory* **3** (1969).

[HOV] Hubbard, J. and Oberste-Vorth, R. Hénon Mappings in the Complex Domain, I: The global topology of dynamical space. *Inst. Hautes Etudes Sci. Publ. Math.* No. 79 (1994), 5–46.

[J] Julia, G. Iteration des Applications Fonctionelles, *J. Math. Pures Appl.* (1918), 47-245.

[LV] Lehto, O. and Virtannen, K.I. *Quasi Conformal Mappings.* Springer-Verlag, Berlin, 1965.

[Ma] Mandelbrot, B. *The Fractal Geometry of Nature*, Freeman & Co., San Francisco, 1982.

[McM] McMullen, C. T. *Complex Dynamics and Renormalization.* Princeton University Press, 1994.

[Mi] Milnor, J. *Dynamics in One Complex Variable.* Vieweg, 2000.

[Mo] Morrey, C. B. On the Solutions of Quasi-Linear Elliptic Partial Differential Equations. *Trans. Amer. Math. Soc.* **43** (1938), 126-166.

[Su] Sullivan, D. Quasiconformal Maps and Dynamical Systems I, Solutions of the Fatou-Julia Problem on Wandering Domains. *Ann. Math.* **122** (1985), 401-418.

[W] Wagoner, J. Realizing Symmetries of a Shift. *Ergodic Theory Dynam. Systems* **8** (1988), 459-481.

Department of Mathematics, Boston University, Boston, MA 02215
E-mail address: paul@bu.edu

Department of Mathematics, Boston University, Boston, MA 02215
E-mail address: bob@bu.edu

Department of Mathematics, Herbert H. Lehman College, CUNY, Bronx, N. Y. 10468
E-mail address: linda@alpha.lehman.cuny.edu

Proceedings of Symposia in Applied Mathematics
Volume **60**, 2004

Multi-Dimensional Symbolic Dynamics

Douglas Lind

CONTENTS

1. Introduction

Dynamics has traditionally been the study of the iterates of a single transformation, modeling the time evolution of a physical system. However, many physical and mathematical systems have other symmetries as well. This leads directly to the study of the joint action of several commuting transformations. This lecture will introduce simple examples of such actions, describe some of their properties and significant theorems, and discuss a few of the many open problems. The main message is that the study of joint actions is much more than a routine generalization of a single transformation, and that genuinely new and deep phenomena occur that are only now being understood.

We will begin by describing the higher-dimensional analogue of a shift of finite type, which consists of all d-dimensional arrays of symbols from a finite alphabet subject to a finite number of local rules or conditions. Such arrays can be shifted in each of the d coordinate directions, giving d commuting invertible transformations, or what amounts to the same thing, an action of the lattice $\mathbb{Z}^d$ of d-tuples of integers. Already a deep distinction arises between $d = 1$, where it is quite easy to describe the space of such arrays, and $d \geqslant 2$, where there is no general algorithm which will decide, given the set of local rules, whether or not the space is empty!

A good example to keep in mind is that of Wang tiles. Imagine a finite set of unit squares, with each square having each of its four edges colored (different edges on the same tile are allowed to have the same color). The corresponding two-dimensional shift of finite type is the set of all tilings of the plane by copies of these squares, where the corners of the squares are aligned with integral lattice points, subject to the local rule that common edges must have the same color. Different collections of Wang tiles lead to quite different shifts of finite type, some of which are well understood and some of which are still mysterious.

Although multi-dimensional shifts of finite type are in general still quite difficult to understand, there is one class of commuting transformations for which a systematic and powerful theory has been recently developed. These are *algebraic* actions, which are commuting automorphisms of compact abelian groups. Here is a simple but very instructive example: consider all two-dimensional arrays of 0's and 1's subject to the condition that, at each site, the sum of the digits at the site itself, the one to the right, and the one above is even. It is not hard to see that this set is a compact abelian group under coordinate-wise operations, and that the horizontal and vertical shifts are commuting group automorphisms. We will see how very concrete examples like this can be thoroughly investigated using tools from harmonic analysis and commutative algebra.

Finally, we will conclude with a taste of open problems, such as Lehmer's conjecture on Mahler measure and its connections with entropy, the computation of information capacity that arises in holographic data storage, and Furstenberg's conjecture about measures simultaneously invariant under several commuting transformations.

We thank Greg Friedman and an anonymous referee for many suggestions to improve an earlier draft of these notes.

2. Multi-dimensional shifts of finite type

Perhaps the most important notion in symbolic dynamics is that of a shift of finite type. As you have already heard in other lectures, this can be described as the set of all bi-infinite sequences of symbols drawn from a finite alphabet, subject to a finite number of local (that is, depending on only finitely many coordinates) rules or conditions that depend only on patterns appearing and not on the particular location in the sequence. For example, the golden mean shift is the set of all bi-infinite sequences of 0's and 1's such that no two 1's occur in adjacent positions.

We can formulate the multi-dimensional version of this idea (replacing sequences with arrays) without much difficulty. However, as we'll soon see, the resulting problems are often much harder, and the general theory of multi-dimensional shifts of finite type is at a much more rudimentary stage.

Start with a finite alphabet $\mathcal{A}$. Often we take $\mathcal{A} = \{0, 1\}$ in our examples. Let $d \geqslant 1$, which will be the dimension of the arrays considered. Most of the time we take $d = 2$, since we can draw 2-dimensional arrays on 2-dimensional paper, and this case illustrates practically all of the new problems and phenomena.

The arrays we'll be considering are indexed by the d-dimensional lattice $\mathbb{Z}^d$. First a little notation for $\mathbb{Z}^d$. We'll use bold-face letters like $\mathbf{k}$, $\mathbf{m}$, and $\mathbf{n}$ for elements of $\mathbb{Z}^d$. Each lattice point is expressed using coordinates as in $\mathbf{n} = (n_1, \dots, n_d)$. The jth standard basic vector is denoted $\mathbf{e}_j = (0, \dots, 0, 1, 0, \dots, 0)$, with a 1 in the jth position. The length of a vector $\mathbf{n}$ in $\mathbb{Z}^d$ is the usual Euclidean length defined by $\|\mathbf{n}\| = \sqrt{n_1^2 + \cdots + n_d^2}$.

We'll use the notation $x = \{x_{\mathbf{n}}\}$ for a typical point in the space, where $x_{\mathbf{n}}$ is the element of $\mathcal{A}$ at location $\mathbf{n} \in \mathbb{Z}^d$. The set of all such arrays will be denoted by $\mathcal{A}^{\mathbb{Z}^d}$. Two points (i.e. arrays) x and y in $\mathcal{A}^{\mathbb{Z}^d}$ are considered "close" if they agree on a large ball around the origin. In other words, x and y are close if there is a large number R such that $x_{\mathbf{n}} = y_{\mathbf{n}}$ for all $\mathbf{n}$ with $\|\mathbf{n}\| < R$. It's not hard to turn this idea into an explicit metric, and with respect to this metric the space $\mathcal{A}^{\mathbb{Z}^d}$ is compact.

Let W be a finite subset of $\mathbb{Z}^d$, which you should think of as a "window" through which to look at a finite part of an array. Our local conditions or rules will take the following form: there is a finite set of allowed patterns of letters from $\mathcal{A}$ one can see looking through W, and a point $x \in \mathcal{A}^{\mathbb{Z}^d}$ is allowed precisely when looking at its elements in every shift of W you see only allowed patterns.

To make this precise, let $\mathcal{P}$ be an arbitrary subset of $\mathcal{A}^W$, which we'll think of as being the allowed patterns you can see through W. From the finite data d, $\mathcal{A}$, W, and $\mathcal{P}$ we define the *d-dimensional shift of finite type over the alphabet $\mathcal{A}$ based on $\mathcal{P}$* to be

$$X_{\mathcal{P}} = \{\, x \in \mathcal{A}^{\mathbb{Z}^d} : x|_{W+\mathbf{k}} \in \mathcal{P} \text{ for all } \mathbf{k} \in \mathbb{Z}^d \,\},$$

where $x|_{W+\mathbf{k}}$ denotes the element of $\mathcal{A}^W$ whose $\mathbf{n}$th coordinate is $x_{\mathbf{n}+\mathbf{k}}$ for all $\mathbf{n} \in W$.

Since the conditions defining $X_{\mathcal{P}}$ are local, it's clear that $X_{\mathcal{P}}$ is a closed subset of $\mathcal{A}^{\mathbb{Z}^d}$. That's because if a sequence $x^{(n)}$ of points in $X_{\mathcal{P}}$ converges to $x \in \mathcal{A}^{\mathbb{Z}^d}$, then at every translate of W the coordinates of the $x^{(n)}$ must stabilize to some constant array which is one of the allowed ones in $\mathcal{P}$, and therefore looking at x through this translate of W also will be allowed.

The dynamics gets into the act by using shifts of arrays in each of the d coordinate directions.

An *action β of $\mathbb{Z}^d$ on a space Y* is a map β from $\mathbb{Z}^d$ to the group of invertible maps of Y. The image of $\mathbf{n} \in \mathbb{Z}^d$ is denoted by $\beta^{\mathbf{n}}$, a convenient exponential notation that imitates the usual one of successive composition for $\mathbb{Z}$-actions, i.e., iterates of a single map.

Using this notation, we can define the *$\mathbb{Z}^d$-shift action σ* on $X_{\mathcal{P}}$ by

$$(\sigma^{\mathbf{n}}(x))_{\mathbf{k}} = x_{\mathbf{n}+\mathbf{k}}.$$

Thus $\sigma^{\mathbf{e}_k}$ shifts an array one unit in the $-\mathbf{e}_k$ direction (to see why the shift is $-\mathbf{e}_k$ think of how the graph of $y = f(x+1)$ is related to the graph of $y = f(x)$). Note the essential property that shifts in different coordinate directions *commute*, so that the order doesn't matter. This means that σ is completely specified once we know the d commuting maps $\sigma^{\mathbf{e}_k}$, $k = 1, 2, \ldots, d$.

Note that when $d = 1$ we have defined the traditional shift of finite type from symbolic dynamics, as described in the introductory article by Susan Williams.

3. Examples

Here we give some simple examples of 2-dimensional shifts of finite type.

EXAMPLE 3.1. (*The full shift*) The set of all arrays $\mathcal{A}^{\mathbb{Z}^2}$ is clearly a shift of finite type over $\mathcal{A}$. We could take W to be the empty set, or alternatively, just let $\mathcal{P} = \mathcal{A}^W$ be the set of all possible patterns for any given W. When $\mathcal{A} = \{0, 1\}$, you can think of $X_{\mathcal{P}}$ as being the set of all possible arrays of 0's and 1's in the plane.

EXAMPLE 3.2. (*Golden mean shift*) Let $\mathcal{A} = \{0, 1\}$, and X be the set of arrays in $\mathcal{A}^{\mathbb{Z}^2}$ such that there are never two 1's adjacent either horizontally or vertically. This is called the *2-dimensional golden mean shift*, by analogy with the 1-dimensional golden mean shift from symbolic dynamics (the name ultimately comes from the fact that the entropy of the 1-dimensional system is the logarithm of the golden mean, i.e., $\log[(1 + \sqrt{5})/2]$). This example is sometimes also referred to as the *hard sphere model*, since the presence of a particle (represented by a "1") at a given site prevents the presence of a particle at neighboring sites.

We can give a presentation of this shift of finite type by using an L-shaped window $W = \{(0, 0), (1, 0), (0, 1)\}$, and letting $\mathcal{P}$ be the set of all patterns in $\mathcal{A}^W$ for which there are never two adjacent 1's either horizontally or vertically. Thus if the pattern has a 0 at $(0, 0)$, then anything is allowed in the other two places, while if there is a 1 at $(0, 0)$, then the other two symbols must be 0. A little reflection shows that $X_{\mathcal{P}}$ is exactly the space defined in the previous paragraph. A typical point from this space is shown in Figure 3.1, together with some sample translates of W to illustrate the allowed patterns.

$$
\begin{matrix}
 & & \vdots & & \\
 & 0\ 0\ 1\ 0\ 0\ 1\ 0\ 1\ 0\ 0\ 0\ 0\ 1\ 0\ 0\ 1\ 0\ 0\ 0\ 1 & \\
 & 0\ 0\ 0\ 0\ 1\ 0\ 0\ 0\ 0\ 1\ 0\ 0\ 0\ 1\ 0\ 0\ 0\ 0\ 1\ 0 & \\
 & 0\ 1\ 0\ 1\ 0\ 0\ 0\ 1\ 0\ 0\ 0\ 0\ 1\ 0\ 0\ 1\ 0\ 0\ 0\ 0 & \\
 & 0\ 0\ 1\ 0\ 0\ 0\ 1\ 0\ 0\ 1\ 0\ 1\ 0\ 0\ 0\ 0\ 0\ 1\ 0\ 1 & \\
 & 1\ 0\ 0\ 0\ 1\ 0\ 0\ 1\ 0\ 0\ 1\ 0\ 0\ 1\ 0\ 1\ 0\ 0\ 1\ 0 & \\
 & 0\ 0\ 0\ 1\ 0\ 0\ 0\ 0\ 1\ 0\ 0\ 0\ 1\ 0\ 1\ 0\ 0\ 1\ 0\ 0 & \\
\cdots & 0\ 1\ 0\ 0\ 0\ 1\ 0\ 0\ 0\ 0\ 0\ 1\ 0\ 0\ 0\ 1\ 0\ 0\ 0\ 1 & \cdots \\
 & 1\ 0\ 0\ 0\ 1\ 0\ 0\ 1\ 0\ 1\ 0\ 0\ 1\ 0\ 1\ 0\ 0\ 0\ 0\ 0 & \\
 & 0\ 0\ 1\ 0\ 0\ 1\ 0\ 0\ 0\ 0\ 0\ 1\ 0\ 1\ 0\ 0\ 1\ 0\ 1\ 0 & \\
 & 1\ 0\ 0\ 0\ 1\ 0\ 0\ 0\ 0\ 0\ 1\ 0\ 0\ 0\ 0\ 1\ 0\ 1\ 0\ 0 & \\
 & 0\ 1\ 0\ 1\ 0\ 0\ 0\ 1\ 0\ 0\ 0\ 1\ 0\ 1\ 0\ 0\ 0\ 0\ 0\ 0 & \\
 & 0\ 0\ 0\ 0\ 1\ 0\ 1\ 0\ 0\ 1\ 0\ 0\ 1\ 0\ 0\ 1\ 0\ 1\ 0\ 1 & \\
 & 1\ 0\ 1\ 0\ 0\ 0\ 0\ 1\ 0\ 0\ 0\ 1\ 0\ 0\ 1\ 0\ 1\ 0\ 0\ 0 & \\
 & & \vdots & &
\end{matrix}
$$

FIGURE 3.1. Typical point from the golden mean shift

EXAMPLE 3.3. (*Ledrappier's example*) We again let $\mathcal{A} = \{0, 1\}$ and use the same L-shaped window $W = \{(0, 0), (1, 0), (0, 1)\}$. However, this time we require that the number of 1's appearing in the window be even. Thus we can either see all 0's through the window, or a pattern with exactly two 1's, but no other. A typical point from this 2-dimensional shift of finite type is depicted in Figure 3.2, with four windows that show all allowed patterns.

There is a somewhat more algebraic way to describe this shift of finite type, one which will be used in much more detail in Section 6. Consider $\mathcal{A}$ as a finite group, where group operations are taken modulo 2. A more standard way to write this is to take $\mathcal{A} = \mathbb{Z}/2\mathbb{Z}$. The condition for a point in $(\mathbb{Z}/2\mathbb{Z})^{\mathbb{Z}^2}$ to be in $X_{\mathcal{P}}$ is that at each coordinate location, when we add up the value at this location, the one to the right, and the one above, we should always get 0 in the group $\mathbb{Z}/2\mathbb{Z}$. This is

```
                         ⋮
    1 0 0 0 1 0 1 0 0 0 1 1 1 1 0 0 1 1 0 1
    1 0 0 0 0 1 1 0 0 0 0 1 0 1 0 0 0 1 0 0
    0 1 1 1 1 1 0 1 1 1 1 1 0 0 1 1 1 1 0 0
    1 1 0 1 0 1 0 0 1 0 1 0 1 1 1 0 1 0 1 1
    1 0 1 1 0 0 1 1 1 0 0 1 1 0 1 0 0 1 1 0
    1 0 0 1 0 0 0 1 0 1 1 1 0 1 1 0 0 0 1 0
··· 0 1 1 1 0 0 0 0 1 1 0 1 0 0 1 0 0 0 0 1 ···
    1 1 0 1 0 0 0 0 0 1 0 0 1 1 1 0 0 0 0 0
    1 0 1 1 0 0 0 0 0 0 1 1 1 0 1 0 0 0 0 0
    0 1 1 0 1 1 1 1 1 1 1 0 1 0 0 1 1 1 1 1
    0 0 1 0 0 1 0 1 0 1 0 1 1 0 0 0 1 0 1 0
    1 1 1 0 0 0 1 1 0 0 1 1 0 1 1 1 1 0 0 1
    1 0 1 0 0 0 0 1 0 0 0 1 0 0 1 0 1 0 0 0
                         ⋮
```

FIGURE 3.2. Typical point from Ledrappier's example

an algebraic condition, and it's easy to see that if both x and y satisfy it, then so do $x+y$ and $-x$ (actually, since here $-1=1$ in $\mathbb{Z}/2\mathbb{Z}$, the second part is trivial). It follows that $X_{\mathcal{P}}$ is not only a compact space, but it also a compact *group*, with coordinate-wise group operations. Furthermore, the elements of the shift action are actually group isomorphisms, i.e., $\sigma^{\mathbf{n}}(x+y)=\sigma^{\mathbf{n}}(x)+\sigma^{\mathbf{n}}(y)$. This is an important example of an *algebraic* $\mathbb{Z}^d$*-action*, which we'll be considering in Section 6. For such actions, many of the most troubling questions have very good answers.

Although Examples 3.2 and 3.3 use the same window and the same number of allowed patterns, their dynamical behavior is quite different. For example, as we'll see later there is a precise quantitative sense in which Example 3.2 has many more overall patterns occurring in it than does Example 3.3.

EXAMPLE 3.4. (*Wang tiles*) This example uses a more geometric flavor to define 2-dimensional shifts of finite type, which you might run into tiling your bathroom wall. Start with a finite set $\mathcal{T}$ of closed 1×1 squares, called *tiles*, such that each tile has each of its four edges colored with various colors (different edges on the same tile could be colored the same). Such a collection $\mathcal{T}$ is called a set of *Wang tiles* (see [**35**] and [**29**]). Loosely speaking, we want to tile all of $\mathbb{R}^2$ with copies of tiles from $\mathcal{T}$ such that the colors of overlapping edges match up correctly.

To make this description more concrete, for each tile $T\in\mathcal{T}$ let $r(T)$ and $l(T)$ denote the colors of its right and left edges, respectively, and similarly $t(T)$ and $b(T)$ the colors of its top and bottom edges. We are allowed to use an unlimited number of copies of each of the tiles. A *Wang tiling* w of $\mathbb{R}^2$ by $\mathcal{T}$ is a covering of $\mathbb{R}^2$ by translates of tiles from $\mathcal{T}$ (no rotations allowed) such that

(1) the corners of each tile are aligned on lattice points in $\mathbb{Z}^2$,
(2) two tiles of w whose edges overlap must have those edges colored with the same color (so that if T and T' are tiles with T immediately to the left of T', then $r(T)=l(T')$, and similarly if T is immediately below T', then $t(T)=b(T')$).

Using $\mathcal{T}$ as the alphabet, we identify a tiling w with a point x in $\mathcal{T}^{\mathbb{Z}^2}$ by letting $x_{\mathbf{n}}$ be the tile in w whose lower left corner is at $\mathbf{n}$. The set $X_{\mathcal{T}}$ of all such points is clearly a shift of finite type, called the *Wang shift* for tile set $\mathcal{T}$.

Here is an explicit example of a 2-dimensional Wang shift: let $\mathcal{T}_D$ be the set of Wang tiles

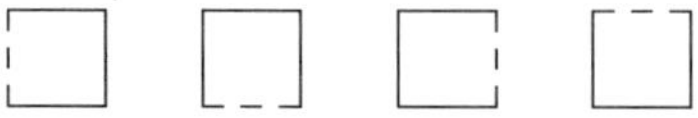

with two colors for edges, indicated by solid or broken lines. The following picture shows a partial Wang tiling of $\mathbb{R}^2$ by $\mathcal{T}_D$ and explains the name "domino tiling" for such a tiling: two tiles meeting along an edge colored with a broken line form a single vertical or horizontal "domino."

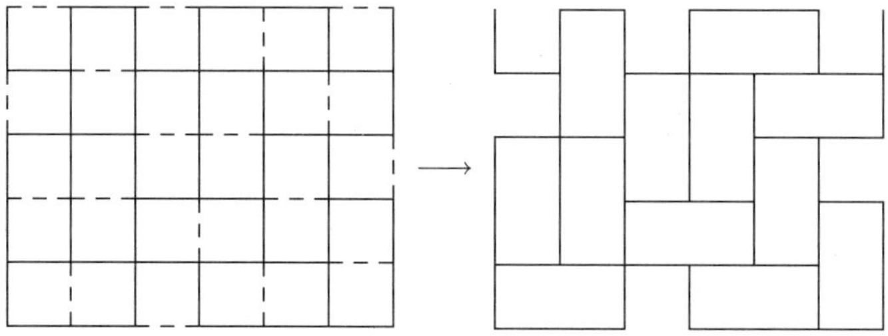

The Wang shift $X_{\mathcal{T}_D}$ for the tile set $\mathcal{T}_D$ is called the *domino shift* or *dimer shift*. It is one of the few non-algebraic 2-dimensional shifts of finite type which is reasonably well-understood. For more on the domino shift, see [**2**], [**7**], [**34**], and especially [**9**], where its mixing properties are investigated.

The definition of Wang tiling has obvious analogues for any $d \geqslant 1$: a collection of *d-dimensional Wang tiles* is a finite set $\mathcal{T}$ of d-dimensional cubes with colored $(d-1)$-faces. The corresponding *Wang shift* $X_{\mathcal{T}} \subset \mathcal{T}^{\mathbb{Z}^d}$ consists of all coverings of $\mathbb{R}^d$ by translates of copies of elements of $\mathcal{T}$ with non-overlapping interiors satisfying the analogues of conditions (1) and (2) above.

EXAMPLE 3.5. (*Cellular automata*) Cellular automata are the ultimate discretization of physical models: space, time, and measured quantities are all discrete. The amazing thing about them is that with very simple local rules it is possible to generate behavior of striking complexity. In fact, a recent book by Stephen Wolfram [**36**] argues that such models provide a better basis of understanding physical reality than the more traditional approach of differential equations or other continuous models.

To define a cellular automata, start with a finite alphabet $\mathcal{A}$. For each point $x \in \mathcal{A}^{\mathbb{Z}}$ we will generate a new point in the same space using a *local rule* at each site. A *local rule of range* r is simply a function $\Phi\colon \mathcal{A}^{2r+1} \to \mathcal{A}$, where $\mathcal{A}^{2r+1}$ is the set of all $(2r+1)$-tuples of symbols from $\mathcal{A}$. For $x \in \mathcal{A}^{\mathbb{Z}}$ we form its image $y = \phi(x)$ at site n by looking at the coordinates of x from $n-r$ to $n+r$, applying Φ to this $(2r+1)$-tuple, and putting the result as the nth coordinate of y. In symbols,

$$y_n = [\phi(x)]_n = \Phi(x_{n-r}, x_{n-r+1}, \dots, x_{n+r-1}, x_{n+r}).$$

We can, of course, repeat this operation, and obtain $\phi(y) = \phi^2(x)$, and so on. Much of the appeal of the subject lies in representing a point in $\mathcal{A}^{\mathbb{Z}}$ as an infinite line of colored squares, one color for each symbol in $\mathcal{A}$. The image point is then appended immediately below this line, and its image immediately below, and so on, creating a 2-dimensional array of colored squares. The local rule shows that this array obeys a 2-dimensional finite-type constraint: the window is $W = \{(-r,0),(-r+1,0),\dots,(r-1,0),(r,0),(0,-1)\}$, and a pattern in $\mathcal{A}^W$ is allowed exactly when the symbol at $(0,-1)$ is the image under Φ of the coordinates at $(-r,0)$ through $(r,0)$. Strictly speaking, this is an array indexed by $\mathbb{Z}\times\{\dots,-3,-2,-1,0\}$ rather than $\mathbb{Z}^2$, so it is not quite a 2-dimensional shift of finite type. Nevertheless, most of what we discuss applies to such examples.

When $\mathcal{A} = \{0,1\}$ we can use a white square to represent a "0" and a black square to represent a "1". To give a concrete example, let $r = 1$, and $\Phi\colon \{0,1\}^3 \to \{0,1\}$ be defined by $\Phi(000) = \Phi(001) = \Phi(010) = \Phi(111) = 1$, while $\Phi(011) = \Phi(100) = \Phi(101) = \Phi(110) = 0$. Starting with a random point of length 100, Figure 3.3 depicts the resulting evolution of this rule. This illustration uses periodic boundary conditions to compute ϕ at both ends of each horizontal line of symbols.

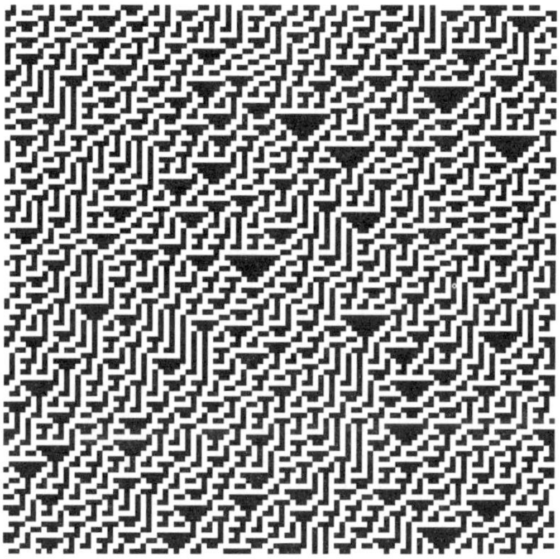

FIGURE 3.3. Evolution of a cellular automaton

The introductory article by Susan Williams describes how linear algebra and matrix theory provide powerful tools for dealing with one-dimensional shifts of finite type. Attempts such as [**24**] to generalize these matrix methods to higher-dimensional shifts of finite type met with only limited success. However, there is a class of two-dimensional shifts of finite type called *textile systems*, introduced by Nasu [**26**], where graphs and matrices play a similarly useful role.

4. The Swamp of Undecidability

Suppose you are handed the finite data to specify a d-dimensional shift of finite type. This means you are given the alphabet $\mathcal{A}$, the finite window W, and the finite set $\mathcal{P}$ of allowed patterns in $\mathcal{A}^W$. These define the resulting shift of finite type $X_{\mathcal{P}}$. Immediately an embarrassing question comes up: Are there *any points at all* in $X_{\mathcal{P}}$?

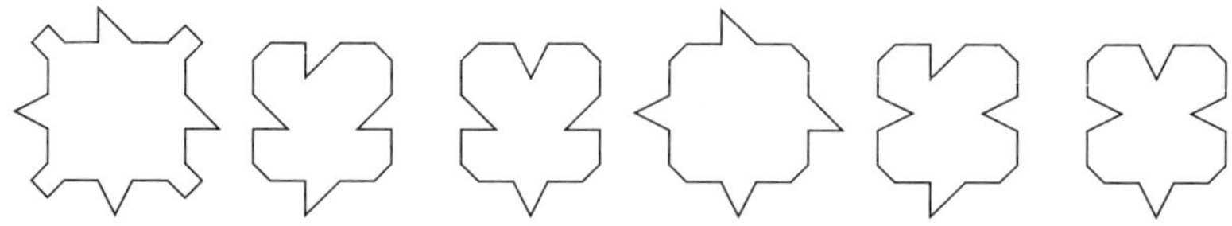

FIGURE 4.1. Robinson tiles

When $d = 1$ it is not hard to decide this: once you start building up a point using allowed patterns, you can extend the future and the past independently, so that they never interfere. This is the *Markov* property for such shifts, and is a key reason for their nice (and tractable) behavior.

However, for $d \geqslant 2$ this becomes a *much* more complicated question, for now what you try to do in one region of $\mathbb{Z}^2$ in building up an allowed point might very well affect what you can do elsewhere. As a simple example, suppose you tile a finite region of the plane with dominoes. How can you tell whether or not you can continue to tile the entire plane starting with this pattern?

The best resolution to this problem would be an algorithm, or computer program, which would accept as input the finite data describing a shift of finite type, and in a finite amount of time say definitely whether or not there are any points in the space. Unfortunately, it is now known that no such computer program exists: the existence problem for higher-dimensional shifts of finite type is *undecidable.* Let's try to understand exactly what this means. Certainly there are some sets of finite data for which it's clear that there are lots of points in the resulting shifts of finite type (the examples in the previous section provide a few). However, there is no uniform procedure, guaranteed to stop in a finite amount of time, that will take *every* instance of this situation and produce a definite answer. This discovery of Berger [**1**] in 1966 means that it's probably hopeless to strive for a completely general theory of higher-dimensional shifts of finite type.

Subsequently Robinson [**29**] greatly simplified and extended Berger's ideas. He found a finite set of tiles for which the extension problem is also undecidable: there is no general algorithm which answers the question in a finite amount of time whether a given finite configuration can be extended to an infinite tiling of the plane.

A traditional way to investigate the dynamics of a single transformation is to study its periodic points, namely those points whose orbit under the transformation is finite. Such a definition of periodicity also makes sense for d-dimensional shifts of finite type. A point $x \in X_{\mathcal{P}}$ is *periodic* if the set of images $\{\sigma^{\mathbf{n}}(x) : \mathbf{n} \in \mathbb{Z}^d\}$ is finite. When $d = 1$ a nonempty shift of finite type always contains periodic points. Unfortunately, this no longer holds when $d \geqslant 2$.

EXAMPLE 4.1. Consider the set of six polygonal tiles from Robinson's paper [**29**] shown in Figure 4.1, each of which should be thought of as a 1×1 square with various bumps and dents. Let $\mathcal{T}$ be the set of all tiles obtained from these six tiles by allowing horizontal and vertical reflections as well as rotations by multiples of 90°. This generates a tiling space $X_{\mathcal{T}}$ much like Wang tiles, where here the bumps and dents take the role of the colors. Clearly $X_{\mathcal{T}}$ is a 2-dimensional shift of finite type. In [**29**], Robinson shows that $X_{\mathcal{T}}$ is uncountable, and yet it has no periodic points at all. Roughly speaking, his argument shows that any allowed tiling of $\mathbb{R}^2$ must have a hierarchical structure that prevents periodic behavior.

The fact that none of these three basic questions, (1) the existence of points, (2) the extension of finite configurations, (3) the existence of periodic points, can be decided by a finite procedure is what I call the "Swamp of Undecidability." It's a place you don't want to go. Fortunately, there are classes of higher-dimensional shifts of finite type for which these questions can be answered, the nicest being that of algebraic $\mathbb{Z}^d$-actions, which will be treated in Section 6.

For more about tilings and shifts of finite type without periodic points, you can read paper by Mozes [**25**] and the introductory book by Radin [**27**]. The pinwheel tiling of the plane featured in the latter is being used to stunning effect as the exterior of the Federal Square building project in Melbourne, Australia (you can see for yourself at `www.federalsquare.com.au`). The article by Robby Robinson has much more to say about tilings and their dynamical properties.

5. Conjugacy and invariants

A basic question throughout all of mathematics is whether two objects, perhaps defined in quite different ways, are really the "same." For example, given two groups given by generators and relations, are the groups isomorphic? For this example being the "same" means being isomorphic as groups. What does it mean for two d-dimensional shifts of finite type to be the "same"?

The answer lies in the notion of sliding block code, which we've already seen when we considered cellular automata. To describe this idea, let $\mathcal{A}$ and $\mathcal{B}$ be two alphabets, possibly the same. Let F be a finite subset of $\mathbb{Z}^d$, and $\Phi : \mathcal{A}^F \to \mathcal{B}$ be a function defined for each configuration of shape F with symbols in $\mathcal{A}$ and having values in $\mathcal{B}$. Define a map $\phi\colon \mathcal{A}^{\mathbb{Z}^d} \to \mathcal{B}^{\mathbb{Z}^d}$ by the equation

$$\phi(x)_{\mathbf{n}} = \Phi(x|_{F+\mathbf{n}}),$$

where $x|_{F+\mathbf{n}}$ denotes the restriction of x to the set $F+\mathbf{n}$. In other words, to compute the $\mathbf{n}$th coordinate of the image, we slide F by $\mathbf{n}$, look at what we see there, and then use Φ on this pattern to compute a symbol from $\mathcal{B}$. Such a map is called a *sliding block code*. When $d = 1$ this is none other than a cellular automaton map as described in Example 3.5.

Suppose that $\mathcal{P}$ is a set of patterns from alphabet $\mathcal{A}$ and $\mathcal{Q}$ is a set of patterns from alphabet $\mathcal{B}$. We'll say that the two shifts of finite type $X_{\mathcal{P}}$ and $X_{\mathcal{Q}}$ are *topologically conjugate* (or simply *conjugate*) if there is a sliding block code $\phi\colon X_{\mathcal{P}} \to X_{\mathcal{Q}}$ that is both one-to-one and onto. Topological conjugacy turns out to be symmetric, in the sense that if $\phi\colon X_{\mathcal{P}} \to X_{\mathcal{Q}}$ is a sliding block code, then the point map $\phi^{-1}\colon X_{\mathcal{Q}} \to X_{\mathcal{P}}$ is also a sliding block code, and hence a conjugacy, although this takes a bit of theory to prove. For us, conjugate shifts of finite type are the "same". We'll also say that $X_{\mathcal{Q}}$ is a *factor* of $X_{\mathcal{P}}$ if there is a sliding block code $\phi\colon X_{\mathcal{P}} \to X_{\mathcal{Q}}$ that is onto (but not necessarily one-to-one). Although the term "factor" would seem to imply something like a direct summand, and although "quotient" would be more consistent terminology, the use of "factor" has been so firmly established that it can't be changed.

As a simple example, let $\mathcal{A} = \{0,1\}$, $\mathcal{B} = \{a,b\}$, $F = \{\mathbf{0}\}$, and $\Phi(0) = a$, $\Phi(1) = b$. This map takes an element of the full 2-shift from Example 3.1 and replaces 0's with a's and 1's with b's. Clearly the image is for all practical purposes the "same" as the original full shift. A good example of a factor map is obtained using the 1-dimensional full 2-shift for both spaces, and the sliding block code simply adding (mod 2) the coordinate at a given site to the one immediately to its right. This gives an onto map from the 2-shift to itself which is everywhere two-to-one. A two-dimensional analogue of this factor map is the sliding block code $\phi\colon \{0,1\}^{\mathbb{Z}^2} \to \{0,1\}^{\mathbb{Z}^2}$ given by the local rule

$$\phi(x)_{(m,n)} = x_{(m,n)} + x_{(m+1,n)} + x_{(m,n+1)} \pmod 2,$$

which is onto and everywhere four-to-one.

As one might expect from our trip through the Swamp of Undecidability, there is no general procedure for determining with an algorithm whether two arbitrary d-dimensional shifts of finite type are conjugate or not. Even when $d = 1$, although there has been substantial progress on this problem, a complete answer is still out of reach. Jack Wagoner's Short Course chapter treats this conjugacy problem for 1-dimensional shifts of finite type in detail.

When confronted with situations where one is not able to immediately determine whether two objects are the same, mathematicians rely on the use of *invariants.* They attach to each object some mathematical structure, say a group or a ring or a vector space or a number, such that if two objects are the same, then the resulting structures are also the same. A simple case is that of dimension for vector spaces: every vector space has a whole number attached to it, its dimension, and isomorphic vector spaces have the same dimension. Here the invariant is decisive: if two vector spaces have the same dimension, they must be isomorphic. Generally speaking, invariants give imperfect pictures of the objects they represent. However, we do know one thing for sure: if two object have different values of the invariant, then they are definitely *not* the same.

To be useful, an invariant should be sensitive (i.e. detect the difference between many pairs of objects) and yet computable. This tension plays out in various ways, and the search for good invariants is a basic activity in many branches of mathematics.

Here's a simple example of an invariant for a shift of finite type: the number of fixed points, i.e., the number of points that remain fixed under each element of the shift. Clearly conjugate shifts have the same number of fixed points. So if these numbers differ, we know at once the shifts are not conjugate. For instance, the full 2-shift in Example 3.1 has precisely two fixed points, the one consisting of all 0's and the other consisting of all 1's. However, Ledrappier's example or the golden mean shifts from Examples 3.3 and 3.2 each have only one fixed point, namely the one of all 0's. So neither of these can be conjugate to the full 2-shift.

This idea can be expanded to periodic points. If x is a periodic point for a shift of finite type σ, then $\{\mathbf{n} \in \mathbb{Z}^d : \sigma^{\mathbf{n}}(x) = x\}$ is a subgroup of $\mathbb{Z}^d$ with finite index (it's quotient group has the same number of elements as the orbit of x). Thus for every finite-index subgroup $\Lambda \subset \mathbb{Z}^d$ we can introduce the invariant that is the number of points in the space such that $\Lambda = \{\mathbf{n} \in \mathbb{Z}^d : \sigma^{\mathbf{n}}(x) = x\}$. This gives infinitely many invariants for shifts of finite type. However, as we have seen in the last section, it is not decidable whether any of these numbers is nonzero!

There is another quantitative invariant called *entropy*. This counts the growth rate of the number of patterns one can see in a square of side n. To simplify the discussion, let's assume $d = 2$, and consider an $n \times n$ square Q_n. Let r_n denote the number of possible patterns from $\mathcal{A}^{Q_n}$ that appear as you consider all points in the shift of finite type $X_{\mathcal{P}}$. Typically, the numbers r_n will grow roughly like $\exp[cn^2]$, and the value of the constant c is called the *entropy* of the shift of finite type. More formally, we define the entropy of the map $\sigma_{\mathcal{P}}$ to be

$$h(\sigma_{\mathcal{P}}) = \limsup_{n\to\infty} \frac{1}{n^2} \log r_n,$$

where log denotes the natural logarithm.

Perhaps it is not immediately obvious, but a little reflection will show that conjugate shifts of finite type must have the same entropy. Roughly speaking, squares in one shift can be coded by squares in the other that are larger by a fixed amount. This leads to certain "edge effects" in the estimates, which become negligible compared to the n^2 contributions from the squares as n becomes large. In addition, a factor of a shift of finite type has entropy that is less than or equal to that of the domain shift.

EXAMPLE 5.1. (1) For the full 2-shift in Example 3.1, every square pattern is allowed, so that $r_n = 2^{n^2}$. Thus the entropy is $\log 2$.

(2) Consider Ledrappier's Example 3.3. A line segment of coordinates determines the segment of coordinates immediately above it having the same left endpoint and length shorter by one, using the L-shaped rule. Thus a line segment of 0's and 1's of length $2n$ on the base determines a square of coordinates of side n. It follows that $r_n \leqslant 2^{2n}$, and dividing $\log r_n$ by n^2 kills it off, so that the entropy of this example is zero.

(3) The entropy of the domino shift in Example 3.4 was computed by Kasteleyn [**13**] to be

$$\frac{1}{4}\int_0^1 \int_0^1 \log(4 - 2\cos 2\pi s - 2\cos 2\pi t)\, ds\, dt \approx 0.291561.$$

(4) For the golden mean shift Example 3.2, the value of entropy has been computed numerically to be between 0.4074 and 0.4078 [**6**], but there is no known closed form expression for it.

REMARK 5.2. Recently scientists at IBM and elsewhere have been working on holographic data storage, in which two-dimensional arrays of bits are stored in a hologram, rather than the ususal one-dimensional strings of bits. The two-dimensional arrays that can be used are typically subject to the sorts of finite type constraints we have been discussing. Given such constraints, the entropy of the resulting shift of finite type provides an upper bound on the information density such a holographic device can attain. Therefore it is of great practical interest to compute, even approximately, the entropies of multi-dimensional shifts of finite type.

6. Algebraic $\mathbb{Z}^d$-actions

One way to avoid the Swamp of Undecidability is to put some additional structure on the shifts of finite type considered. The most successful application of this idea has been to ask that the space also be a *group*, so that there is a notion of addition of points in the space, and the elements of the shift act as group automorphisms.

We've already seen a few examples of this. If we consider $\mathcal{A} = \{0, 1\}$ as the group $\mathbb{Z}/2\mathbb{Z}$ with addition modulo 2, the full 2-shift in Example 3.1 is a group under coordinate-wise operations: $(x+y)_{\mathbf{n}} = x_{\mathbf{n}} + y_{\mathbf{n}}$. Since shifting and then adding is the same as adding and then shifting, for each $\mathbf{k} \in \mathbb{Z}^2$ we have that $\sigma^{\mathbf{k}}(x+y) = \sigma^{\mathbf{k}}(x) + \sigma^{\mathbf{k}}(y)$, so that elements of the shift action are group homomorphisms, in fact isomorphisms since they are clearly one-to-one and onto. Here the shift space is really the product group $(\mathbb{Z}/2\mathbb{Z})^{\mathbb{Z}^2}$.

Ledrappier's Example 3.3 is also a group, since the defining relation using the L-shaped window is additive: if both x and y satisfy it, then so does $x+y$. This group is a closed subgroup of the full 2-shift. In fact, given any finite "shape" in $\mathbb{Z}^2$, one can define a similar sort of shift, by requiring that when you add up all coordinates within any translate of this shape you get 0 (mod 2). Each of these is a group, in addition to being a shift of finite type. What makes these examples tractable is that there is a huge amount of machinery available, in the form of abstract harmonic analysis, commutative algebra, and number theory. We'll be using some of this machinery, but even if you're not familiar with it you should be able to get a feel for what's going on.

What we have been seeing in the last few examples are cases of an *algebraic $\mathbb{Z}^d$-action*. This is defined as follows. Let X be a compact abelian group, and let $\operatorname{aut}(X)$ denote the group of all continuous group automorphisms of X. For example, let X be the n-dimensional torus $\mathbb{T}^n$, where $\mathbb{T} = \mathbb{R}/\mathbb{Z}$. Then $\operatorname{aut}(X) = GL(n,\mathbb{Z})$, the group of all $n \times n$ integral matrices with determinant ± 1. An *algebraic $\mathbb{Z}^d$-action* on X is a map $\alpha\colon \mathbb{Z}^d \to \operatorname{aut}(X)$. The image of $\mathbf{n} \in \mathbb{Z}^d$ under α is denoted using the exponential notation $\alpha^{\mathbf{n}}$, so that $\alpha^{\mathbf{m}+\mathbf{n}} = \alpha^{\mathbf{m}} \circ \alpha^{\mathbf{n}}$ and $\alpha^{\mathbf{0}}$ is the identity automorphism on X.

A good argument can be made that compactness is the analyst's version of finiteness (see the illuminating article by Hewitt [**12**] on this idea). With this in mind, we enlarge our idea of shift of finite type to include the case when the alphabet is a compact group. For many reasons, such as the Hilbert Basis Theorem, this continues to be a finitely-described object. For instance, the $\mathbb{Z}^2$ shift action on $\mathbb{T}^{\mathbb{Z}^2}$ gives an algebraic $\mathbb{Z}^2$-action with alphabet $\mathcal{A} = \mathbb{T}$. The analogue of Ledrappier's example is then the subgroup specified by requiring that the three coordinates in any translate of the L-shaped window add up to 0, but now in the group $\mathbb{T}$ rather than $\mathbb{Z}/2\mathbb{Z}$.

Let $\mathbb{S} = \{z \in \mathbb{C} : |z| = 1\}$ denote the multiplicative version of the circle group. A *character* χ of the compact abelian group X is a continuous group homomorphism $\chi\colon X \to \mathbb{S}$. The set $\widehat{X}$ of all characters of X is also a group under pointwise operations, called the *Pontryagin dual group* of X. For instance, if $X = \mathbb{T}$ then $\widehat{X} \cong \mathbb{Z}$ since every character of $\mathbb{T}$ has the form $t \mapsto e^{2\pi i n t}$ for some $n \in \mathbb{Z}$. An automorphism ϕ of X induces a dual automorphism $\widehat{\phi}$ of $\widehat{X}$ using the formula $[\widehat{\phi}(\chi)](x) = \chi[\phi(x)]$. Therefore if α is an algebraic $\mathbb{Z}^d$-action on X, the duals of the d generating automorphisms $\alpha^{\mathbf{e}_j}$ induce d commuting automorphisms of the dual group $\widehat{X}$. Let $R_d = \mathbb{Z}[u_1^{\pm 1}, \dots, u_d^{\pm 1}]$ denote the ring of Laurent polynomials in d commuting variables with integer coefficients. We can then make $\widehat{X}$ into an R_d-module by the formula $u_j \cdot \chi = \widehat{\alpha^{\mathbf{e}_j}}(\chi)$. Thus an algebraic $\mathbb{Z}^d$-action dualizes to an R_d-module.

This process can be reversed! If M is an R_d-module, then its dual group $X_M = \widehat{M}$ is a compact abelian group. Each u_j gives an automorphism of M via the action $m \mapsto u_j \cdot m$, and therefore these automorphisms dualize to d commuting automorphisms of X_M. We call the resulting $\mathbb{Z}^d$-action α_M.

The basic result of Pontryagin duality says that taking the dual group of the dual group gets us back to the group we started with. For example, the dual of $\mathbb{T}$ is $\mathbb{Z}$, and the dual of $\mathbb{Z}$ is $\mathbb{T}$. This is analogous to duality in finite-dimensional vector spaces, where the double dual of a vector space is isomorphic in a natural way to the original vector space. For us, this means that duality establishes a one-to-one correspondence between the set of all algebraic $\mathbb{Z}^d$-actions α and the set of all R_d modules M, with an R_d-module M corresponding to the algebraic $\mathbb{Z}^d$-action α_M on X_M. This is the crucial link that allows the use of powerful tools from algebra and algebraic geometry.

The book by Schmidt [**33**] gives a detailed account of algebraic $\mathbb{Z}^d$-actions, and is loaded with interesting examples together with much of the basic theory. We'll just touch on a few of these developments here.

Recall that an ideal $\mathfrak{p}$ in R_d is called *prime* if whenever $f \cdot g \in \mathfrak{p}$, then either $f \in \mathfrak{p}$ or $g \in \mathfrak{p}$ (or both). An R_d-module M is *Noetherian* if every increasing chain of submodules stabilizes at some finite stage. Since R_d is a Noetherian ring, equivalently an R_d-module is Noetherian if it is finitely generated over R_d. A prime ideal $\mathfrak{p}$ is *associated to* M if there is an $m \in M$ such that $\mathfrak{p} = \{f \in R_d : f \cdot m = 0\}$. If M is Noetherian, then the set $\operatorname{asc}(M)$ of prime ideals associated to M is always finite. For example, $\operatorname{asc}(R_d/\mathfrak{p}) = \{\mathfrak{p}\}$ for a prime ideal $\mathfrak{p}$. Finding the prime ideals associated to a given Noetherian R_d-module is roughly like factoring a polynomial into a finite number of irreducible polynomials, but generally it is much harder.

The *basic building blocks* of algebraic $\mathbb{Z}^d$-actions are those of the form $\alpha_{R_d/\mathfrak{p}}$, where $\mathfrak{p}$ is a prime ideal in R_d. In a precise sense, every algebraic $\mathbb{Z}^d$-action can be built up from these basic building blocks using extensions. Thus if one knows answers to dynamical questions for these building blocks, one can generally answer these questions for arbitrary algebraic $\mathbb{Z}^d$-actions.

Let me give two examples of this. Every compact abelian group X carries a unique translation invariant measure μ, called Haar measure, such that $\mu(X) = 1$. On the circle group this is simply the usual Lebesgue measure. Also, a group automorphism automatically preserves Haar measure, so that every algebraic $\mathbb{Z}^d$-action does likewise. It is therefore natural to ask when the action is *ergodic*, namely when does it happen that if a subset E of X is invariant under every element of the action, then one can conclude that E has Haar measure either 0 or 1? The answer is that α_M is ergodic if and only if $\alpha_{R_d/\mathfrak{p}}$ is ergodic for each of the associated prime ideals $\mathfrak{p} \in \operatorname{asc}(M)$.

But when is one of the basic building blocks $\alpha_{R_d/\mathfrak{p}}$ ergodic? The answer involves the *variety of* $\mathfrak{p}$, defined as follows. Let $\mathbb{C}^\times = \mathbb{C} \setminus \{0\}$ (we omit 0 since we are dealing with Laurent polynomials). Then define

$$V(\mathfrak{p}) = \{\mathbf{z} = (z_1, \dots, z_d) \in (\mathbb{C}^\times)^d : f(\mathbf{z}) = 0 \text{ for all } f \in \mathfrak{p}\}.$$

An element $\mathbf{z} \in (\mathbb{C}^\times)^d$ is called a *unit root* if each entry is a root of unity in $\mathbb{C}$. Then the result is that $\alpha_{R_d/\mathfrak{p}}$ is ergodic if and only if $V(\mathfrak{p})$ does not consist of a finite number of unit roots. This is analogous to the fact that an automorphism of a torus is ergodic if and only if the automorphism has no eigenvalue that is a root of unity.

Another basic property that an action may have is expansiveness. This is a certain finiteness condition with many facets and implications. An algebraic $\mathbb{Z}^d$-action α on X is *expansive* if there is a neighborhood U of the identity 0_X such that

$$\bigcap_{\mathbf{n}\in\mathbb{Z}^d} \alpha^{\mathbf{n}}(U) = \{0_X\}.$$

This says that if two points remain close under all iterates of the action, then they must be equal. For instance, an automorphism of the torus is expansive if and only if none of its eigenvalues has absolute value one, i.e., the automorphism is *hyperbolic.*

Again the reduction to the basic building blocks follows the same pattern: An algebraic $\mathbb{Z}^d$-action α_M is expansive if and only if M is Noetherian and $\alpha_{R_d/\mathfrak{p}}$ is expansive for each of the finite number of prime ideals $\mathfrak{p} \in \operatorname{asc}(M)$. For the basic building blocks, $\alpha_{R_d/\mathfrak{p}}$ is expansive if and only if

$$V(\mathfrak{p}) \cap \mathbb{S}^d = \varnothing,$$

i.e., no point in the variety of $\mathfrak{p}$ has all of its coordinates with absolute value one. This is the d-dimensional version of hyperbolicity for toral automorphisms we mentioned earlier.

The book [**33**] develops a detailed "dynamics/algebra dictionary," translating dynamical properties into corresponding algebraic properties via duality. Sometimes tools already developed in algebra can be used to answer these questions, but sometimes this leads to the development of new tools as well. This topic has been undergoing vigorous development over the past fifteen years, and there is still much more to be learned.

7. Entropy and Lehmer's conjecture

In Section 5 we defined an invariant called entropy for shifts of finite type. This is actually a special case of the definition for topological actions of $\mathbb{Z}^d$ on compact metric spaces. It thus makes sense to ask for the value of entropy for the algebraic $\mathbb{Z}^d$-actions from the previous section. The answer turns out to be quite interesting, and closely connected to an unresolved conjecture made by D. H. Lehmer nearly seventy years ago.

To define entropy, let (Y, ρ) be a compact metric space. A topological $\mathbb{Z}^d$-action β on Y is a homomorphism $\beta\colon \mathbb{Z}^d \to \operatorname{homeo}(Y)$ from $\mathbb{Z}^d$ to the group of homeomorphisms of Y. Thus both d-dimensional shifts of finite type and algebraic $\mathbb{Z}^d$-actions are examples of topological $\mathbb{Z}^d$-actions.

Let $Q_n = \{0, 1, \dots, n-1\}^d$ denote the cube in $\mathbb{Z}^d$ of side n. Fix an $\epsilon > 0$. A set $F \subset Y$ is said to (n, ϵ) *span* Y if for every $y \in Y$ there is a $z \in F$ such that $\rho(\beta^{\mathbf{k}}(y), \beta^{\mathbf{k}}(z)) < \epsilon$ for all $\mathbf{k} \in Q_n$. Let $r_n(\epsilon, \beta)$ denote the size of the (n, ϵ) spanning set with the fewest number of elements (this is a finite number since the space is compact). Define

$$r(\epsilon, \beta) = \limsup_{n\to\infty} \frac{1}{n^d} \log r_n(\epsilon, \beta),$$

and define the topological entropy of β to be

$$h(\beta) = \lim_{\epsilon\to 0} r(\epsilon, \beta).$$

Suppose that A is an automorphism of the n-dimensional torus $\mathbb{T}^n$. Thus A is given by an $n \times n$ integral matrix whose determinant is ± 1. For a set to be (n, ϵ) spanning for A (as a $\mathbb{Z}$-action), it should be roughly $\epsilon|\lambda_j|^{-n}$ dense in each eigenspace for eigenvalue λ_j with $|\lambda_j| > 1$ (the eigenvalues smaller than one in absolute value do not contribute to entropy). Thus in this example

$$r_n(\epsilon, A) \approx \prod_{|\lambda_j|>1} \frac{|\lambda_j|^n}{\epsilon}.$$

Applying the definition of entropy then shows that

$$h(A) = \sum_{|\lambda_j|>1} \log |\lambda_j|.$$

Using Jensen's formula from complex analysis, it is possible to rewrite this formula in a format that makes possible a far-reaching generalization. Let f_A be the characteristic polynomial of A. Then by Jensen's formula [**37**],

$$\int_0^1 \log |f_A(e^{2\pi it})| \, dt = \sum_{|\lambda_j|>1} \log |\lambda_j|.$$

The quantity on the left-hand side has arisen in many contexts, and is the logarithm of the *Mahler measure* of the polynomial f_A. In general, the Mahler measure of a monic polynomial f in one variable is defined to be

$$\mathsf{M}(f) = \exp\left[\int_{\mathbb{S}} \log |f|\right] = \exp\left[\int_0^1 \log |f(e^{2\pi it})| \, dt\right].$$

It is the geometric mean of f over the multiplicative circle. There is also an interpretation of $\mathsf{M}(f)$ as the "L^0 norm of f", in the sense that $\mathsf{M}(f) = \lim_{p\to 0} \|f\|_p$, where $\|f\|_p$ is the usual p-norm $(\int_{\mathbb{S}} |f|^p)^{1/p}$ (see [**30**, Exer. 3.5(d)]). Incidentally, the use of "measure" is only meant to convey that $\mathsf{M}(f)$ "measures" the complexity of a polynomial f, and has nothing to do with the more typical use of the word as a function on subsets of a measure space.

There is a several-variable version of Mahler measure, defined in an analogous fashion. For $f \in R_d$, the ring of Laurent polynomials in d commuting variables, we define its Mahler measure to be

$$\begin{aligned}\mathsf{M}(f) &= \exp\left[\int_{\mathbb{S}^d} \log |f|\right] \\ &= \exp\left[\int_0^1 \cdots \int_0^1 \log |f(e^{2\pi it_1}, \dots, e^{2\pi it_d})| \, dt_1 \dots dt_d\right]\end{aligned}$$

This quantity was introduced by Mahler [**23**] when studying how to measure the "complexity" of polynomials. Define the *length* of a polynomial

$$f(u) = c_0 u^n + c_1 u^{n-1} + \cdots + c_n \quad \text{to be} \quad \ell(f) = \sum_{k=0}^{n} |c_k|.$$

Clearly $\ell(f \cdot g) \leqslant \ell(f)\ell(g)$, but can the length of a product of two polynomials collapse to be much smaller than the product of the lengths? The example

$$(u-1)(u^n + u^{n-1} + \cdots + u + 1) = u^{n+1} - 1$$

shows that the length of the product can be substantially smaller then the product of the lengths. What Mahler needed for his work on transcendental numbers (numbers like π that aren't the roots of polynomials with integer coefficients) is that some sort of reverse inequality holds. He showed that

$$\ell(f \cdot g) \geqslant 2^{-\deg f - \deg g}\, \ell(f)\ell(g).$$

The crucial fact he used in his proof is that $\mathsf{M}(f \cdot g) = \mathsf{M}(f)\mathsf{M}(g)$.

We can use Mahler measure to compute entropy for the basic building blocks for algebraic $\mathbb{Z}^d$-actions. Let $\mathfrak{p}$ be a prime ideal in R_d. Then by [**21**],

$$h(\alpha_{R_d/\mathfrak{p}}) = \begin{cases} \log \mathsf{M}(f) & \text{if } \mathfrak{p} = \langle f \rangle \neq 0 \text{ is a principal ideal,} \\ 0 & \text{if } \mathfrak{p} \text{ is not principal,} \\ \infty & \text{if } \mathfrak{p} = 0. \end{cases} \tag{7.1}$$

EXAMPLE 7.1. (1) Ledrappier's Example 3.3 has the form $\alpha_{R_d/\mathfrak{p}}$ with $\mathfrak{p} = \langle 2, 1 + u_1 + u_2 \rangle$, which is a prime ideal. However, this is not a principal ideal, so that according to (7.1), $h(\alpha_{R_d/\mathfrak{p}}) = 0$, in agreement with Example 5.1(2).

(2) Consider the $\mathbb{T}$ version of Ledrappier's example, in which the alphabet is $\mathbb{T}$ and the L-shaped window condition is the only one. This example has the form $\alpha_{R_d/\mathfrak{p}}$ with $\mathfrak{p} = \langle 1 + u_1 + u_2 \rangle$, which is also prime. However, here the ideal is principal. Thus

$$h(\alpha_{R_d/\mathfrak{p}}) = \log \mathsf{M}(1 + u_1 + u_2) \approx 0.3032.$$

There is a way to express this quantity as

$$L(2, \chi_3) = \sum_{n=1}^{\infty} \frac{\chi_3(n)}{n^2},$$

where $\chi_3(n)$ equals 0, 1, or -1, depending on whether n is congruent to 0, 1, or 2 (mod 3). Glimmerings of a deep connection between Mahler measures of polynomials and values of L-functions are starting to appear.

(3) For the $d = 3$ version of Ledrappier for $\mathbb{T}$, we can similarly compute its entropy to be

$$\log \mathsf{M}(1 + u_1 + u_2 + u_3) = \frac{7\zeta(3)}{2\pi^2},$$

where $\zeta(3)$ is the value of the usual Riemann zeta function at 3.

How are the values of Mahler measure, or equivalently entropy of algebraic $\mathbb{Z}^d$-actions distributed? Can the entropies of such an action be positive but arbitrarily close to zero? Surprisingly, the answer is still unknown, and goes under the name of Lehmer's Conjecture. One way of stating this conjecture, perhaps more appropriately called Lehmer's Problem, is whether given $\epsilon > 0$ there is a monic polynomial f with integer coefficients such that $1 < \mathsf{M}(f) < 1 + \epsilon$. This amounts to asking whether the product of the roots outside the unit circle of such a polynomial can have absolute value very close to 1. Despite much serious work on this problem, it has remained open for seventy years. The smallest known value of Mahler measure greater than one is about 1.17, corresponding to the polynomial

$$u^{10} + u^9 - u^7 - u^6 - u^5 - u^4 - u^3 + u + 1$$

found by Lehmer himself. This polynomial has one root outside the unit circle, one inside, and the rest on the unit circle. Roots of irreducible integer polynomials outside the unit circle for which the other roots are either on or inside the unit circle (and with some roots definitely on the circle) are called Salem numbers, and arise in a variety of mathematical contexts. For (much) more about Salem numbers and Mahler measure, see the article [**3**] by Boyd.

One indication of the relationship of Lehmer's Conjecture with algebraic dynamics is the result in [**19**] that there exists an ergodic automorphism of the infinite torus $\mathbb{T}^{\mathbb{Z}}$ with finite entropy if and only if the answer to Lehmer's question is "yes"!

8. Rigidity and Furstenberg's conjecture

Suppose that the mapping $\phi\colon \mathbb{R} \to \mathbb{R}$ satisfies *Cauchy's functional equation*, namely $\phi(x+y) = \phi(x) + \phi(y)$ for all $x, y \in \mathbb{R}$. What can we say about ϕ? With no further assumptions, it is possible to cook up exceedingly strange ϕ's using a *Hamel basis for* $\mathbb{R}$, which uses the Axiom of Choice to find a basis for $\mathbb{R}$ as a vector space over the rationals $\mathbb{Q}$. But these solutions are so strange (for example, they cannot be measurable functions) that you wouldn't want to meet them in a dark alley! In fact, as soon as you impose even slightly stronger assumptions on ϕ, then the only solutions are the obvious linear ones $\phi(x) = cx$ for some constant c. For example, if ϕ is bounded on an interval, or even measurable, then it must be linear. This is a simple example of a rigidity phenomenon: a group homomorphism from $\mathbb{R}$ to $\mathbb{R}$ that is assumed to be measurable must in fact be algebraic (linear). Rigidity phenomena are becoming increasingly important as we investigate actions of more complicated groups.

Roughly speaking, rigidity for actions of groups like $\mathbb{Z}^d$ says that the orbits of such actions are so intertwined and woven together that mappings which preserve the orbit structure and are measurable must automatically have much stronger properties as well. Another variant of this idea is that measures invariant under the action of a sufficiently complicated group action must be few and classifiable. This means there is a large distinction between actions of $\mathbb{Z}$ (where there are typically a huge number of invariant measures), and rigid actions of more complicated groups (where there are conjecturally very few invariant measures).

The prototype problem for rigidity is the following question, essentially first raised by Furstenberg [**11**] (actually the question is only hinted at there; the first explicit statement seems to be in the paper of Lyons [**22**]). Consider the maps $\phi_n\colon \mathbb{T} \to \mathbb{T}$ given by $\phi_n(t) = nt$. A measure μ on $\mathbb{T}$ is *invariant* for ϕ_n if $\mu(\phi_n^{-1}(E)) = \mu(E)$ for every measurable set $E \subset \mathbb{T}$ (checking this for intervals E is actually enough). For example, Lebesgue measure is invariant for each ϕ_n, since the inverse image of an interval consists of n intervals, each exactly an nth as long as the original interval. A measure is called *continuous* if it gives mass zero to every point.

CONJECTURE 8.1 (Furstenberg). *The only continuous measure on* $\mathbb{T}$ *that is simultaneously invariant under* ϕ_2 *and* ϕ_3 *is Lebesgue measure.*

We remark that there are billions of continuous measures invariant under just ϕ_2, and billions others invariant under ϕ_3. The substance of this conjecture is that demanding joint invariance under both of these maps drastically cuts down on the possibilities, in fact to just one.

If we make some assumptions on the measure (for example, if one of ϕ_2 or ϕ_3 has positive entropy), then Furstenberg's conjecture is known to be true [**31**]. In addition, there has been recent progress for analogues of this conjecture for toral automorphisms [**15**]. Nevertheless, our inability to settle this and other problems points out our ignorance, and suggests that there are yet very exciting new ideas yet to be discovered!

References

[1] R. Berger, The undecidability of the Domino Problem, Mem. Amer. Math. Soc. **66** (1966).

[2] Robert Burton and Robin Pemantle, Local characteristics, entropy and limit theorems for spanning trees and domino tilings via transfer-impedances, Ann. Probab. **21** (1993), 1329–1371.

[3] David Boyd, Speculations concerning the range of Mahler's measure, Canad. Math. Bull. **24** (1981), 453–469.

[4] David Boyd, Mahler's measure and special values of L-functions, Experiment. Math. **7** (1998), 37–82.

[5] R. Burton and J. Steif, Some 2-D symbolic dynamical systems: entropy and mixing, in: Ergodic Theory of $\mathbb{Z}^d$-Actions, ed. M. Pollicott and K. Schmidt, London Math. Soc. Lecture Notes **228**, 297–305, Cambridge Univ. Press, 1996.

[6] Neil J. Calkin and Herbert S.Wilf, The number of independent sets in a grid graph, SIAM J. Discrete Math. **11** (1998), no. 1, 54–60

[7] C. Cohn, N. Elkies and J. Propp, Local statistics for random domino tilings of the Aztec diamond, Duke Math. J. **85** (1996), 117–166.

[8] Christian Deninger, Delign periods and mixed motives, K-theory and the entropy of certain $\mathbb{Z}^n$-actions, J. Amer. Math. Soc. **10** (1997), 259–281.

[9] Manfred Einsiedler, Fundamental cocycles and tiling spaces, Ergodic Th. & Dynam. Syst. **21** (2001), 777-800.

[10] Manfred Einsiedler and Klaus Schmidt, Markov partitions and homoclinic points of algebraic $\mathbb{Z}^d$-actions, Proc. Steklov Inst. Math. **216** (1997), 259–279.

[11] H. Furstenberg, Disjointness in ergodic theory, minimal sets, and a problem in Diophantine approximation. Math. Systems Theory **1** (1967), 1–49.

[12] Edwin Hewitt, The rôle of compactness in analysis, Amer. Math. Monthly **67** (1960), 499–516.

[13] P.W. Kasteleyn, The statistics of dimers on a lattice. I, Phys. D **27** (1961), 1209–1225.

[14] A. Katok, S. Katok, and K. Schmidt, Rigidity of measurable structure for algebraic actions of higher-rank abelian groups, Inventiones Math., to appear.

[15] A. Katok and R. Spatzier, Invariant measures for higher-rank hyperbolic abelian actions, Ergodic Th. & Dynam. Syst. **16** (1996), 751–778.

[16] B. Kitchens and K. Schmidt, Isomorphism rigidity of irreducible algebraic $\mathbb{Z}^d$-actions, Inventiones Math. **142** (2000), 559–577.

[17] S. Lang, *Algebra*, 2nd ed., Addison-Wesley, New York, 1984.

[18] F. Ledrappier, Un champ markovian peut être d'entropie nulle et mélangeant, C. R. Acad. Sci. Paris, Ser. A **2807** (1978), 561–562.

[19] Douglas Lind, Ergodic automorphisms of the infinite torus are Bernoulli, Israel J. Math **17** (1974), 162–168.

[20] Douglas Lind and Brian Marcus, *An Introduction to Symbolic Dynamics and Coding*, Cambridge, New York, 1995.

[21] Douglas Lind, Klaus Schmidt, and Thomas Ward, Mahler measure and entropy for commuting automorphisms of compact groups, Inventiones Math. **101** (1990), 593–629.

[22] Russell Lyons, On measures simultaneously 2- and 3-invariant, Israel J. Math. **61** (1988), 219–224.

[23] K. Mahler, On some inequalities for polynomials in several variables, J. London Math. Soc. **37** (1962), 341–344.

[24] Nelson Markley and Michael Paul, Matrix subshifts for $\mathbb{Z}^\nu$ symbolic dynamics, Proc. London Math. Soc. (3) **43** (1981), 251–272.

[25] Shahar Mozes, Aperiodic tilings, Invent. Math. **128** (1997), 603–611.

[26] Masakazu Nasu, *Textile Systems for Endomorphisms and Automorphisms of the Shift*, Memoirs of Amer. Math. Soc. **546**, Providence, 1995.

[27] Charles Radin, *Miles of Tiles*, American Math. Soc., Providence, 1999.

[28] Marina Ratner, Rigidity of horocycle flows, Annals of Math. **115** (1982), 597–614.

[29] R.M. Robinson, Undecidability and nonperiodicity for tilings of the plane, Invent. Math. **12** (1971), 177–209.
[30] W. Rudin, *Real and Complex Analysis*, McGraw-Hill, New York, 1966.
[31] Daniel Rudolph, $\times 2$ and $\times 3$ invariant measures and entropy, Ergodic Th. & Dynam. Syst. **10** (1990), 395–406.
[32] Klaus Schmidt, *Algebraic Ideas in Ergodic Theory*, CBMS Regional Conference Series in Math. **76**, Amer. Math. Soc., Providence, 1990.
[33] Klaus Schmidt, *Dynamical Systems of Algebraic Origin*, Birkhäuser Verlag, Basel-Berlin-Boston, 1995
[34] Klaus Schmidt, Tilings, fundamental cocycles and fundamental groups of symbolic $\mathbb{Z}^d$-actions, Ergod. Th. & Dynam. Sys. **18** (1998), 1473–1525.
[35] H. Wang, Proving theorems by pattern recognition II, AT&T Bell Labs. Tech. J. **40** (1961), 1–41.
[36] Stephen Wolfram, *A New Kind of Science*, Wolfram Science, 2002.
[37] R. M. Young, On Jensen's formula and $\int_0^{2\pi} \log|1-e^{i\theta}|\,d\theta$, Amer. Math. Monthly **93** (1986), 44-45.

Douglas Lind, Department of Mathematics, Box 354350, University of Washington, Seattle, WA 98195

E-mail address: `lind@math.washington.edu`

Proceedings of Symposia in Applied Mathematics
Volume **60**, 2004

Symbolic Dynamics and Tilings of $\mathbb{R}^d$

E. Arthur Robinson, Jr.

ABSTRACT. Aperiodic tilings of Euclidean space can profitably be studied from the point of view of dynamical systems theory. This study takes place *via* a kind of dynamical system called a *tiling dynamical system*.

1. Introduction

In this chapter we study tilings of Euclidean space from the point of view of dynamical systems theory, and in particular, symbolic dynamics. Our goal is to show that these two subjects share many common themes and that they can make useful contributions to each other. The tilings we study are tilings of $\mathbb{R}^d$ by translations of a finite number of basic tile types called "prototiles". A good general reference on tilings is [**GS87**]. The link between tilings and dynamics will be established using a kind of dynamical system called a *tiling dynamical system*, first described by Dan Rudolph [**Rud88**], [**Rud89**]. The parts of the theory we concentrate on here are the parts most closely related to symbolic dynamics. Interestingly, these also tend to be the parts related to the theory of quasicrystals.

A quasicrystal is a solid which, like a crystal, has a regular enough atomic structure to produce sharp spots in its X-ray diffraction patterns, but unlike a crystal, has an aperiodic atomic structure. Because of this aperiodicity, quasicrystals can have "symmetries" forbidden to ordinary crystals, and these can be observed in their X-ray diffraction patterns. The first quasicrystals were discovered in 1984 by physicists at NIST (see [**SBGC84**]) who observed a diffraction pattern with 5-fold rotational symmetry. For a good mathematical introduction to quasicrystals see [**Sen95**].

The theory of quasicrystals is tied up with some earlier work on tiling problems in mathematical logic ([**Wan61**], [**Brg66**]). Central to this circle of ideas is the concept of an *aperiodic* set of prototiles. One of the most interesting aperiodic sets, which anticipated the discovery of quasicrystals, is the set of Penrose tiles, discovered in the early 1970s by Roger Penrose [**Pen74**]. Penrose tilings play a central role in the theory of tiling dynamical systems because they lie at the crossroad of the three main methods for constructing examples: local matching rules, tiling substitutions, and the projection method. As we will see, the tiling spaces constructed

Key words and phrases. Aperiodic tilings, symbolic dynamics, quasicrystals.

by these methods are analogous to three well known types of symbolic dynamical systems: finite type shifts, substitution systems and Sturmian systems.

The connection between tilings and symbolic dynamics goes beyond the analogies discussed above. Since tilings are (typically) multi-dimensional, tiling dynamics is part of the theory of multi-dimensional dynamical systems. We will show below that one can embed the entire theory of $\mathbb{Z}^d$ symbolic dynamics (the subject of Doug Lind's chapter in this volume) into the theory of tiling dynamical systems. It turns out that much of the complication inherent in multi-dimensional symbolic dynamics (what Lind calls "the swamp of undecidability") is closely related to the existence of aperiodic prototile sets.

Finally, one can view tiling dynamical systems as a new *type* of symbolic dynamical system. Since tilings are geometric objects, the groups that act naturally on them are continuous rather than discrete (i.e., $\mathbb{R}^d$ versus $\mathbb{Z}^d$). Because of this, one needs to define a new kind of compact metric space to replace the shift spaces studied in classical symbolic dynamics. We call this space a *tiling space*. Even in the one dimensional case (i.e., for flows) tiling spaces provide a new point of view.

In the first part of this chapter we carefully set up the basic theory of tiling dynamical systems and give complete proofs of the main results. In later sections, we switch to survey mode, giving references to access the relevant literature. Of course there are many topics we can not cover in such a short chapter.

These notes are based on a AMS Short Course presented by the author at the 2002 Joint Mathematics Meeting in San Diego, California. The author wishes to thank Tsuda College in Tokyo, Japan and the University of Utrecht, The Netherlands, where earlier versions of this course were presented. My thanks to the Natalie Priebe Frank and Cliff Hansen for carefully reading the manuscript and making several helpful suggestions. My thanks also to the referee who suggested several substantial improvements.

2. Basic definitions in tiling theory

2.1. Tiles and tilings. A set $D \subseteq \mathbb{R}^d$, $d \geq 1$, is called a *tile* if it is compact and equal to the closure of its interior. We will always assume that tiles are homeomorphic to topological balls, although in some situations it is useful to allow disconnected tiles. Tiles in $\mathbb{R}$ are closed intervals. Tiles in $\mathbb{R}^2$ are often polygons, but fractal tiles also occur frequently in examples.

A *tiling*[1] x of $\mathbb{R}^d$ is a collection of tiles that *pack* $\mathbb{R}^d$ (any two tiles have pairwise disjoint interiors) and that *cover* $\mathbb{R}^d$ (their union is $\mathbb{R}^d$). Two tiles D_1, D_2 are *equivalent*, denoted $D_1 \sim D_2$, if one is a translation of the other. Equivalence class representatives are called *prototiles*.

DEFINITION 2.1. Let $\mathcal{T}$ be a finite set of inequivalent prototiles in $\mathbb{R}^d$. Let $X_{\mathcal{T}}$ be the set of all tilings of $\mathbb{R}^d$ by translations of the prototiles in $\mathcal{T}$. We refer to $X_{\mathcal{T}}$ as a *full tiling space*.

Broadly speaking, geometry is concerned with properties of objects that are invariant under congruence. Similarly, dynamics is generally concerned with group actions. In this chapter, we will be interested in how groups of rigid motions act on

[1]We use the lower case notation x for a tiling because we want to think of x as a point in a tiling space X.

sets of tilings. Because of this, we will distinguish between congruent tilings in $X_{\mathcal{T}}$ that sit differently in $\mathbb{R}^d$. Of central interest will be the action of $\mathbb{R}^d$ by translation.

DEFINITION 2.2. For $\mathbf{t} \in \mathbb{R}^d$ and $x \in X_{\mathcal{T}}$ let $T^{\mathbf{t}}x \in X_{\mathcal{T}}$ be the tiling of $\mathbb{R}^d$ in which each tile $D \in x$ has been shifted by the vector $-\mathbf{t}$, that is $T^{\mathbf{t}}x = \{D - \mathbf{t} : D \in x\}$. We denote this *translation action* of $\mathbb{R}^d$ on $X_{\mathcal{T}}$ by T.

The primary reason for studying T is that it is related to the long range order properties of the tilings in $X_{\mathcal{T}}$. While such properties are geometric in nature, we will gain access to them through dynamical systems theory.

2.2. Local finiteness. Let $\mathcal{T}$ be a set of prototiles. A *$\mathcal{T}$-patch* y is a finite subset $y \subseteq x$ of a tiling $x \in X_{\mathcal{T}}$ such that the union of tiles in y is connected. This union is called the *support* of y and written $\text{supp}(y)$. The notion of *equivalence* extends to patches, and a set of equivalence class representatives of patches is denoted by $\mathcal{T}^*$. The subset of patches of n tiles, called the *n-patches*, is denoted by $\mathcal{T}^{(n)} \subseteq \mathcal{T}^*$.

We will impose one additional condition, called the *local finiteness condition*, on all tiling spaces $X_{\mathcal{T}}$.

DEFINITION 2.3. A tiling space $X_{\mathcal{T}}$ has *finite local complexity* if $\mathcal{T}^{(2)}$ is finite.

Equivalently, $\mathcal{T}^{(n)}$ is finite for each n. Sometimes the geometry of the tiles themselves will impose the local finiteness condition, but we usually need to add it as an extra assumption. From now on, whenever we write $\mathcal{T}$, $\mathcal{T}^*$ or $X_{\mathcal{T}}$, it will always implicitly include a choice of a finite $\mathcal{T}^{(2)}$. When working with polygonal prototiles in $\mathbb{R}^2$, a common way to achieve local finiteness to assume that all tiles meet *edge-to-edge*.

EXAMPLE 2.4. Consider the set $\mathcal{S}$ consisting of a single 1×1 square prototile. Without any local finiteness condition, fault lines exist in the tilings $x \in X_{\mathcal{S}}$ with a continuum of possible displacements. Imposition of the edge-to-edge condition means that every $x \in X_{\mathcal{S}}$ is a translation of a single *periodic* tiling. See Figure 1.

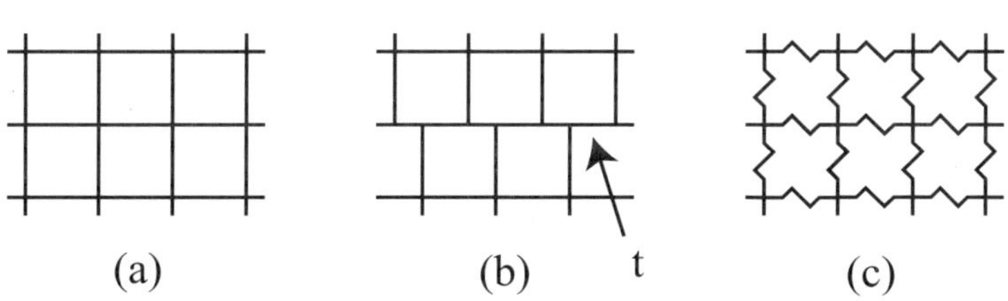

FIGURE 1. (a) Part of an edge-to-edge square tiling. (b) A square tiling with a fault having displacement t. (c) Local finiteness can always be forced geometrically by cutting "keys" on the edges of tiles.

EXAMPLE 2.5. We get more interesting square tiling examples by taking $\mathcal{S}_n$, $n > 1$, to be the set of 1×1 square prototiles *marked* with "colors" $1, 2, \ldots, n$. To do this we also need to modify our notion of equivalence so that differently colored squares are not considered to be equivalent.

Now consider the subset $X_0 \subseteq X_{\mathcal{S}_n}$ consisting of all tilings whose vertices lie on the lattice $\mathbb{Z}^d \subseteq \mathbb{R}^d$ and let T_0 be the restriction of the $\mathbb{R}^d$ shift action T to the subgroup $\mathbb{Z}^d$. It is clear that X_0 is T_0-invariant.

We will see later how this example links tilings to discrete symbolic dynamics.

EXAMPLE 2.6. Fix $n \geq 4$. Let $s = n$ for n odd, and $s = 2n$ for n even. For $0 \leq k < n$ let $\mathbf{v}_k = (\cos(2\pi k/s), \sin(2\pi k/s)) \in \mathbb{R}^2$, i.e., $\mathbf{v}_k$ is a sth root of unity, viewed as a vector in $\mathbb{R}^2$. Let $\mathcal{R}_n$ denote the set of all $\binom{n}{2}$ rhombi with translations of the vectors $\mathbf{v}_k$ as sides. Define $X_{\mathcal{R}_n}$ to be the corresponding edge-to-edge tiling space. Two examples of $x \in X_{\mathcal{R}_5}$ (one with markings) are shown in Figures 3 and 10.

2.3. The tiling topology. As we now show, finite local complexity tiling spaces have particularly nice topological properties. The tiling topology is based on a simple idea: two tilings are close if after a small translation they agree on a large ball around the origin (see [**Rad99**], [**Rob96b**], [**Rud88**], [**Sol97**]). However, the details turn out to be a little subtle.

Given $K \subseteq \mathbb{R}^d$ compact and $x \in X_{\mathcal{T}}$, let $x[[K]]$ denote set of all sub-patches $x' \subseteq x$ such that $K \subseteq \operatorname{supp}(x')$. The smallest such patch is denoted $x[K]$. For $r > 0$ let $B_r = \{\mathbf{t} \in \mathbb{R}^d : ||\mathbf{t}|| < r\}$, where $||\cdot||$ denotes the Euclidean norm on $\mathbb{R}^d$.

LEMMA 2.7. *For $x, y \in X_{\mathcal{T}}$ define*

$$(2.1) \quad d(x,y) = \inf\big(\{\sqrt{2}/2\} \cup \{0 < r < \sqrt{2}/2 : \exists\, x' \in x[[B_{1/r}]], \\ y' \in y[[B_{1/r}]], \text{ with } T^{\mathbf{t}}x' = y' \text{ for some } ||\mathbf{t}|| \leq r\}\big).$$

Then d defines a metric on $X_{\mathcal{T}}$.

We call d the *tiling metric.*

PROOF. We prove only the triangle inequality. Let $0 < d(x,y) = a' \leq d(y,z) = b'$ with $a' + b' < \sqrt{2}/2$. Let $0 < \epsilon < \sqrt{2}/2 - (a' + b')$ and put $a = a' + \epsilon/2$ and $b = b' + \epsilon/2$. Then there are $x' \in x[[B_{1/a}]]$, $y' \in y[[B_{1/a}]]$, $y'' \in y[[B_{1/b}]]$ and $z'' \in z[[B_{1/b}]]$, and also $\mathbf{t}, \mathbf{s} \in \mathbb{R}^d$ with $||\mathbf{t}|| \leq a$ and $||\mathbf{s}|| \leq b$, such that $T^{\mathbf{t}}x' = y'$ and $T^{-\mathbf{s}}z'' = y''$.

Let $y_0 = y' \cap y''$, $x_0 = T^{-\mathbf{t}}y_0 \subseteq x'$ and $z_0 = T^{\mathbf{s}}y_0 \subseteq z'$. Then

$$(2.2) \qquad T^{-(\mathbf{t}+\mathbf{s})}z_0 = x_0 \text{ where } ||\mathbf{t} + \mathbf{s}|| \leq a + b.$$

Letting $c = a + b$, then since $0 < a \leq b < \sqrt{2}/2$,

$$0 \leq \frac{1}{c} = \frac{1}{a+b} \leq \frac{1}{b} - a,$$

and it follows that $B_{1/c} \subseteq (B_{1/b} + \mathbf{t})$. Now $y', y'' \in y[[B_{1/b}]]$ so $x_0 \in x[[B_{1/b} + \mathbf{t}]] \subseteq x[[B_{1/c}]]$.

Combining this with (2.2), we have $d(x,z) \leq a + b = d(x,y) + d(y,z) + \epsilon$, where $\epsilon > 0$ is arbitrarily small. The triangle inequality follows.

□

LEMMA 2.8. *The tiling metric d is complete.*

PROOF. Consider a Cauchy sequence x_n of tilings. Assume $d(x_{n+1}, x_n) > 0$ and let $s_n = d(x_{n+1}, x_n) + 2^{-n}$. By passing to a subsequence, we may assume s_n is decreasing and $\sum_{n=1}^{\infty} s_n < \infty$. it follows from (2.1) that for each n there exists $\mathbf{t}_n \in \mathbb{R}^d$ with $||\mathbf{t}_n|| \leq s_n$ and $x'_n \in x_n[[B_{1/s_n}]]$ such that $T^{\mathbf{t}_n} x'_n \subseteq x'_{n+1}$. Put $\mathbf{r}_n = \sum_{k=n}^{\infty} \mathbf{t}_k$. Then

$$T^{\mathbf{r}_n} x'_n = T^{\mathbf{r}_{n+1}} T^{\mathbf{t}_n} x'_n \subseteq T^{\mathbf{r}_{n+1}} x'_{n+1}.$$

This implies that $T^{\mathbf{r}_n} x'_n$ is an increasing sequence of patches, so we can define a tiling $x = \cup_n T^{\mathbf{r}_n} x'_n$. Finally, $d(x, x_n) \leq \max(||\mathbf{r}_n||, s_n) \to 0$. □

THEOREM 2.9. (Rudolph [**Rud89**]) *Suppose $X_{\mathcal{T}}$ is a finite local complexity tiling space. Then $X_{\mathcal{T}}$ is compact in the tiling metric d. Moreover, the action T of $\mathbb{R}^d$ by translation is on $X_{\mathcal{T}}$ is continuous.*

EXERCISE 1. The proof of compactness amounts to the observation that the local finiteness condition is equivalent to $X_{\mathcal{T}}$ being *totally bounded* (see [**Mun75**]). Fill in the details of this proof and also prove the continuity of T.

EXERCISE 2. Prove the following converse to Theorem 2.9: a translation invariant set X of tilings which is compact in the tiling metric (2.1) must have finite local complexity.

3. Tiling dynamical systems

3.1. Tiling spaces as symbolic dynamical systems. Throughout this chapter, a *dynamical system* will be a pair (X, T) where X is a compact metric space (the *phase space*) and T is a continuous action of a group, usually (but not always) $\mathbb{R}^d$. The study of the topological properties of dynamical systems is called *topological dynamics*. The study of the "statistical properties" of dynamical systems is called *ergodic theory*. An excellent introduction to both topological dynamics and ergodic theory is Walters[2] [**Wal82**].

Symbolic dynamics studies a special kind of dynamical system called a *symbolic dynamical system*. The classical set-up is 1-dimensional, but we describe here the general d-dimensional case (see also the chapter by D. Lind). For the group we take $\mathbb{Z}^d$, and we let $X_n = \{1, \ldots, n\}^{\mathbb{Z}^d}$, $n > 1$, with the product topology. Letting T be the *shift* action of $\mathbb{Z}^d$ on X_n, we obtain a dynamical system (X_n, T) called the *d-dimensional full shift* on *n symbols*. In some ways this example itself is too simple to be interesting, but it has very complicated subsets. A *$\mathbb{Z}^d$-symbolic dynamical system* is defined to be a pair (X, T) where X is a closed T-invariant subset $X \subseteq X_n$ called a *shift space*[3] (see [**LM95**] and [**Que87**]).

DEFINITION 3.1. Let $X_{\mathcal{T}}$ be a full d-dimensional tiling space and let T denote the translation action of $\mathbb{R}^d$. A *tiling space* X is a closed T-invariant subset $X \subseteq X_{\mathcal{T}}$. We call the pair (X, T) a *tiling dynamical system.*

Now we can precisely state our way of thinking of tiling dynamical systems as a new type of symbolic dynamical system. We think of the prototiles $D \in \mathcal{T}$ as the symbols. The full tiling space $X_{\mathcal{T}}$ corresponds to the full shift, and more general tiling spaces correspond to more general shift spaces. Like the product topology,

---

[2]Even though this book concentrates almost exclusively on $\mathbb{Z}$ actions, the theory goes through with very little effort to actions of $\mathbb{Z}^d$ and $\mathbb{R}^d$.

[3]A shift space is also sometimes called a subshift.

the tiling topology is compact and metric, and in both cases closeness corresponds to a good match near the origin. However, in the the case of tilings there is the possibility of a small translation, and since we want this to be continuous, the topology needs to be defined accordingly.

REMARK 3.2. The theory of tiling dynamical systems contains the theory of $\mathbb{Z}^d$ symbolic dynamics. One can show that the space X_0 constructed in Example 2.5 is homeomorphic to the symbolic full shift, and that T_0 implements the shift action on X_0.

3.2. Finite type. Let $X_{\mathcal{T}}$ be a full tiling space and let $\mathcal{F} \subseteq \mathcal{T}^*$. Let $X_{\backslash\mathcal{F}} \subseteq X_{\mathcal{T}}$ be the set of all tilings $x \in X_{\mathcal{T}}$ such that no patch y in x is equivalent to any patch in $\mathcal{F}$. We call such a set $\mathcal{F}$ a set of *forbidden patches.*

One can show that for any $\mathcal{F} \subseteq \mathcal{T}^*$, the set $X_{\backslash\mathcal{F}}$ is a tiling space (i.e., it is closed and T-invariant). Moreover, it is clear that every tiling space $X \subseteq X_{\mathcal{T}}$ is defined by a set $\mathcal{F}$ of forbidden patches. However, the set $\mathcal{F}$ is not unique!

EXERCISE 3. Prove the three statements in the previous paragraph. Hint: See [**LM95**].

In symbolic dynamics, the most important kind of shift space is a finite type shift. The following definition introduces the corresponding idea in tiling theory.

DEFINITION 3.3. A tiling space $X \subseteq X_{\mathcal{T}}$ is called a *finite type tiling space* if there exists a *finite* $\mathcal{F} \subseteq \mathcal{T}^*$ so that $X = X_{\backslash\mathcal{F}}$.

The most common case is $\mathcal{F} \subseteq \mathcal{T}^{(2)}$. This is called a *local matching rule.* It is convenient to formulate this case in terms of the *allowed* 2-patches rather than forbidden ones. To accomplish this, we put $\mathcal{Q} = \mathcal{T}$, let $\mathcal{Q}^{(2)} = \mathcal{T}^{(2)} \backslash \mathcal{F}$, and write $X_{\mathcal{Q}}$ for $X_{\backslash\mathcal{F}}$. Note that imposing a local matching rule really just amounts to strengthening the local finiteness condition. Thus a full tiling space is a kind of finite type tiling space.

EXAMPLE 3.4. (The Penrose tiles) Consider the marked version $\mathcal{P}$ (shown in Figure 2) of the prototiles $\mathcal{R}_5$. The set $\mathcal{P}^{(2)}$ (which defines the matching rules)

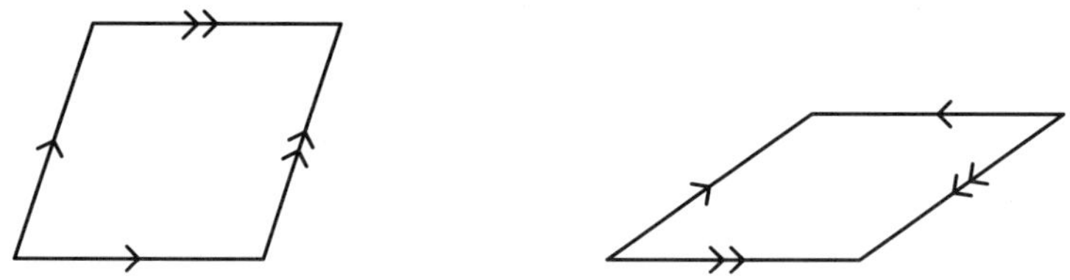

FIGURE 2. The Penrose tiles. The protoset $\mathcal{P}$ consists of the two marked tiles shown, and all rotations so that edges have have angles $2\pi n/10$. In particular, card$(\mathcal{P}) = 20$.

imposes the requirement that the markings on any pair of adjacent tiles must match. (As we will see below, markings are often used for this purpose). We call $\mathcal{P}$, together with the matching rules, the *Penrose tiles.* Tilings $x \in X_{\mathcal{P}}$ are called *Penrose tilings.* Part of a Penrose tiling is shown in Figure 3.

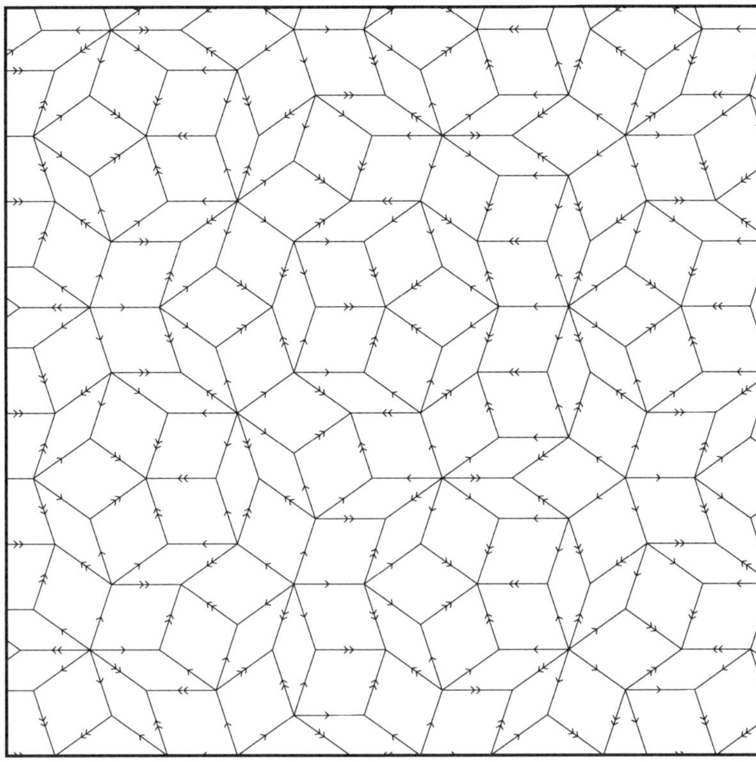

FIGURE 3. A part of a Penrose tiling.

3.3. The Tiling Problem. Suppose we are given a set $\mathcal{T}$ of prototiles and a set $\mathcal{F} \subseteq \mathcal{T}^*$ of forbidden patches. Consider the following problem:

TILING PROBLEM. Is $X_{\setminus\mathcal{F}} \neq \emptyset$?

We begin with a positive result, a version of which appeared in [**Wan61**] (see [**GS87**] for a proof).

EXTENSION THEOREM. *Let $\mathcal{T}$ be a collection of prototiles with a local finiteness condition $\mathcal{T}^{(2)}$ and let $\mathcal{F} \subseteq \mathcal{T}^*$ be a set of forbidden patches. Define $\mathcal{T}^+ \subseteq \mathcal{T}^*$ to be the set of patches that do not contain any forbidden sub-patches. Then $X_{\setminus\mathcal{F}} \neq \emptyset$ if and only for each $r > 0$ there is a patch $y \in \mathcal{T}^+$ with $B_r + \mathbf{t} \subseteq \operatorname{supp}(y)$ for some $\mathbf{t} \in \mathbb{R}^d$.*

The trouble with the Extension Theorem is that it is not constructive. To conclude that $X_{\setminus\mathcal{F}}$ is nonempty one needs to see infinitely many patches in $\mathcal{T}^+$.

This difficulty can be appreciated if one tries to tile the plane manually with Penrose tiles. There are a lot of "dead ends": patches in $y \in \mathcal{P}^+$ that do not belong to $\mathcal{P}^*$. How can we know that $\mathcal{P}^+$ doesn't have some largest patch y? Later, we will give a proof that $X_{\mathcal{P}} \neq \emptyset$, but that proof will require a new idea.

The question of whether the Tiling Problem is *decidable* was raised by Wang [**Wan61**] for the case of marked square tiles $\mathcal{S}_n$, together with a local matching rule. These are now known as Wang tiles.

When $d = 1$ there is an easy algorithm to answer the Tiling Problem: First we draw a graph G with vertex set $\mathcal{T}$ and directed edges $\mathcal{T}^{(2)}$, and let A be the $m \times m$ adjacency matrix for G, where $m = \text{card}(\mathcal{T})$. The entries of A^k give the number of paths of length k in G. If $A^{m+1} \neq \mathbf{0}$ (i.e., not identically zero) then $A^k \neq \mathbf{0}$ for any $k > m$, and we conclude $X_{\mathcal{Q}} \neq \emptyset$.

DEFINITION 3.5. A tiling x of $\mathbb{R}^d$ is called a *periodic tiling* if its *translation group* $\Gamma_x = \{\mathbf{t} \in \mathbb{R}^d : T^{\mathbf{t}}x = x\}$ is a *lattice*: that is a subgroup of $\mathbb{R}^d$ with d linear independent generators. A tiling x is called *aperiodic* if $\Gamma_x = \{0\}$.

In the case $d = 1$ one can easily show that if $X_{\mathcal{T}} \neq \emptyset$ then there is a periodic tiling $x \in X_{\mathcal{T}}$. Wang conjectured [**Wan61**] that the same holds for $d > 1$.

WANG'S CONJECTURE.

(1) *There is an algorithm to decide the tiling problem.*
(2) *Whenever $X_{\mathcal{T}} \neq \emptyset$, there exists a periodic tiling $x \in X_{\mathcal{T}}$.*

Wang proved that (2) implies (1). The argument goes as follows:

First suppose $X_{\mathcal{T}} \neq \emptyset$. Then there exists a periodic $x \in X_{\mathcal{T}}$. For each $n \in \mathbb{N}$, list all tiling patches $y \in \mathcal{T}^+$ with $B_n \subseteq \text{supp}(y)$ and $B_{n+1} \not\subseteq \text{supp}(y)$. We will eventually see a complete period of x. In this case the algorithm will stop and answer "yes".

Now suppose $X_{\mathcal{T}} = \emptyset$. Then it follows from the Extension Theorem that we will eventually find $n \in \mathbb{N}$ so that no $y \in \mathcal{T}^+$ has support containing B_n. In this case the algorithm will stop and answer "no".

But Wang's conjecture turns out to be false! For $d \geq 2$ Berger [**Brg66**] showed that the tiling problem is, in fact, *undecidable*. His solution included the construction of an example of a 2-dimensional finite type tiling space $X_{\mathcal{Q}}$ containing no periodic tilings.

DEFINITION 3.6.

(1) A nonempty tiling space X is called an *aperiodic tiling space* if it contains no periodic tilings (i.e., every $x \in X$ is an aperiodic tiling).
(2) A prototile set $\mathcal{Q}$ with local matching rule $\mathcal{Q}^{(2)}$ is called an *aperiodic prototile set* if $X_{\mathcal{Q}}$ is an aperiodic tiling space.[4]

Berger's original aperiodic prototile set $\mathcal{Q}$ satisfies $\text{card}(\mathcal{Q}) > 50,000$ (see [**GS87**]). Later, Raphael Robinson [**rRob71**] found a simple example with $\text{card}(\mathcal{Q}) = 32$ (a picture of these tiles appears in Lind's chapter of this volume).

3.4. Counting prototiles and the "einstein" problems. The problem of finding small aperiodic sets of prototiles has been a popular one (see [**GS87**] for the history up to 1987). The exact formulation depends on how one counts prototiles. With our notion of equivalence (translation but not more general congruences), the Penrose tiles $\mathcal{P}$ consist of 20 prototiles. Counting this way, the current best example in $\mathbb{R}^2$ is a set $\mathcal{K}$ of Wang tiles due to Kari and Culik ([**Kar96**], [**Cul96**]) with $\text{card}(\mathcal{K}) = 13$. This example is particularly interesting because it is not related to any other known examples.

---

[4]In some literature, the term "aperiodic tiling" is reserved for tilings $x \in X_{\mathcal{Q}}$, where $\mathcal{Q}$ is an aperiodic prototile set.

It is perhaps more natural to allow congruence classes of prototiles to count only once (i.e., to allow rotations of the prototiles). With this system of counting there are just 2 Penrose tiles, and for $\mathbb{R}^2$ this is the best result so far.

The question of whether there exists an aperiodic prototile set consisting of a *single* tile has been named the "einstein problem" by Ludwig Danzer. For $d = 3$ an example of an einstein was discovered by Schmitt and Conway (see [**Sen95**]). However, it tiles in a way that is, in a certain sense, very weakly aperiodic.

There are some partial results on the 2-dimensional einstein problem as well. For any prototile set $\mathcal{T} = \{D\}$, where D is a topological disk, whenever $X_{\mathcal{T}} \neq \emptyset$ there exists a periodic tiling $x \in X_{\mathcal{T}}$ ([**Ken92**], [**Ken93**], [**GBN89**]). In other words, there is no einstein up to translation.

4. Substitution tiling spaces

4.1. Perfect decompositions. Let $L \in Gl(d, \mathbb{R})$ be an *expansive* linear transformation of $\mathbb{R}^d$. Expansive means that every eigenvalue of L lies outside the unit circle. The case $L = \lambda M$, where M is an isometry and $\lambda > 1$ is called a *similarity*.

A *perfect decomposition* (or just a *decomposition*) is a mapping $C : \mathcal{T} \to L^{-1}\mathcal{T}^*$ that (up to equivalence) satisfies the *perfect overlap* condition

$$\operatorname{supp}(C(D)) = \operatorname{supp}(D). \tag{4.1}$$

In a slight abuse of language, a decomposition is called *self-similar* if L is a similarity; in the general case it is called a *self-affine* decomposition.

The mapping $S = LC$ is called a *perfect self-similar* or *self-affine tiling substitution*[5] on $\mathcal{T}$.

In the case $d = 1$ there is no difference between a self-similar and a self-affine substitution, and any tiling substitution can be written $S = \lambda C$, where λ is a positive real number. In the self-similar case when $d = 2$, we can identify $\mathbb{R}^2$ with $\mathbb{C}$. By replacing S with S^2 we can assume L orientation preserving. Then we can regard L as multiplication by $\lambda \in \mathbb{C}$, with $|\lambda| > 1$, so that $S = \lambda C$.

EXAMPLE 4.1. (Polyomino decompositions) The *chair decomposition* C_c on the set $\mathcal{C}$ of four chair prototiles is obtained by taking the decomposition pictured in Figure 4(a) and its four rotations. The *table decomposition* C_t, Figure 4(b), is

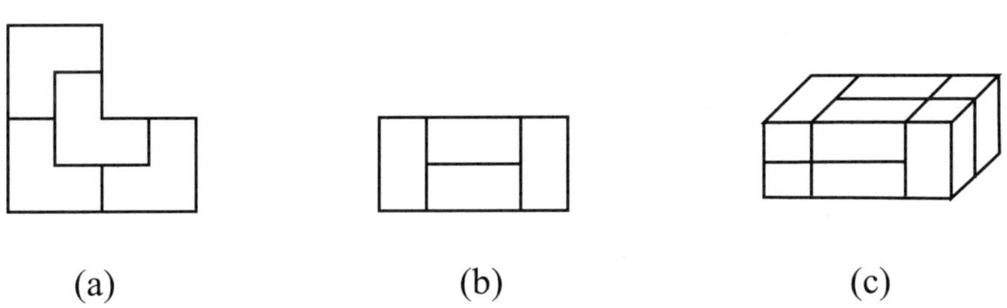

FIGURE 4. (a) The chair, (b) the table and (c) the 3-dimensional table.

defined on the protoset $\mathcal{D}_2$ of two "dimers" in the plane. The 3-dimensional table, Figure 4(c), is defined on the set $\mathcal{D}_3$ of 6 dimers in $\mathbb{R}^3$. The asymmetry of this

[5]This is also sometimes called an *inflation* mapping.

example makes it necessary to keep track of the prototiles' orientations. Many other polyomino examples are easily devised.

In all cases shown in Figure 4, L is a similarity with $M = Id$ and $\lambda = 2$. A non-self-similar polyomino decomposition with $L = \begin{pmatrix} 2 & 0 \\ 0 & 3 \end{pmatrix}$ is shown in Figure 5.

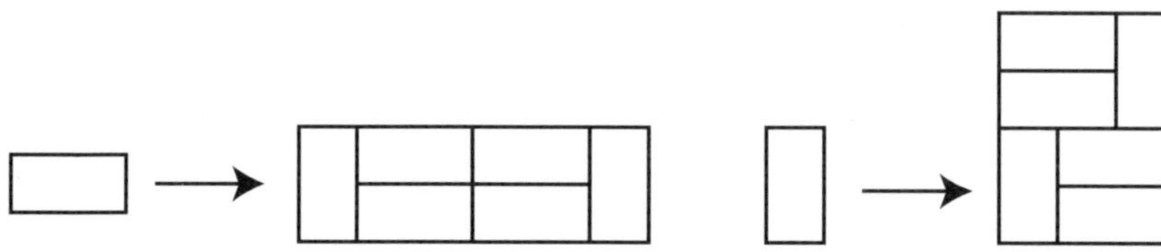

FIGURE 5. The *folding table*: a non-self-similar version of the table. Here we show the decomposition of $L\mathcal{D}_2$ into $\mathcal{D}_2$.

EXAMPLE 4.2 (Raphael Robinson's triangular Penrose tilings). This is a non-polyomino self-similar decomposition C_1 with $\lambda = (1/2)(1+\sqrt{5})$ on a set $\mathcal{P}_1$ of 40 marked triangular prototiles. Two of the prototiles are shown in Figure 6(a). The decomposition C_1 is shown in Figure 6(b). Let $X_{\mathcal{P}_1}$ denote the finite type tiling space corresponding to the usual matching rule that the arrows on adjacent edges must match. We will show below how to use the tiling substitution $S_1 = \lambda C_1$. to prove that $X_{\mathcal{P}_1} \neq \emptyset$.

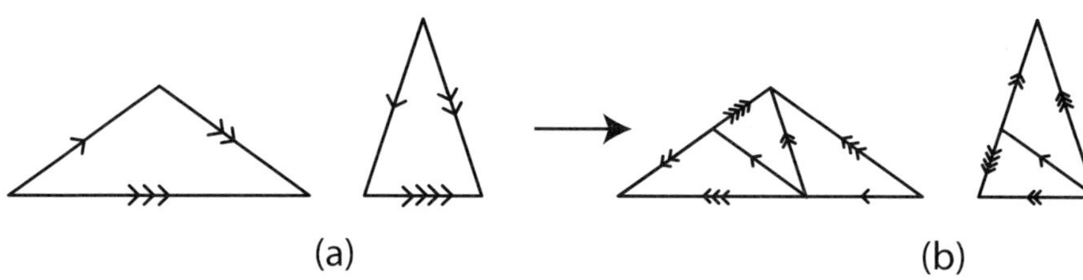

FIGURE 6. The *triangular Penrose tilings*: The protoset $\mathcal{P}_1$ consists of a finite set of rotations of the two tiles shown that is closed under decomposition.

EXAMPLE 4.3 (The pinwheel tilings). The self-similar decomposition shown in Figure 7 has been studied extensively by Radin (see [**Rad94**]). Up to rotation and reflection it has a single prototile. In the decomposition, one copy of the prototile is rotated by an angle $\theta = \arctan\frac{1}{2}$, so that $\frac{\theta}{2\pi}$ is irrational. Such a rotation has infinite order and hence there is no finite prototile set invariant under this decomposition. We can get around this difficulty by modifying the definition of equivalence and the tiling metric d to allow rotations as well as translations.

The next two examples don't quite satisfy (4.1), but can nevertheless easily be accommodated. We refer to them as *imperfect decompositions* and the corresponding substitutions as *imperfect tiling substitutions* (see Definition 5.25).

EXAMPLE 4.4. (The Penrose decomposition). This decomposition applies to the marked prototile set $\mathcal{P}$ of Example 3.4. Since $C(D) \supseteq D$, this is an imperfect decomposition.

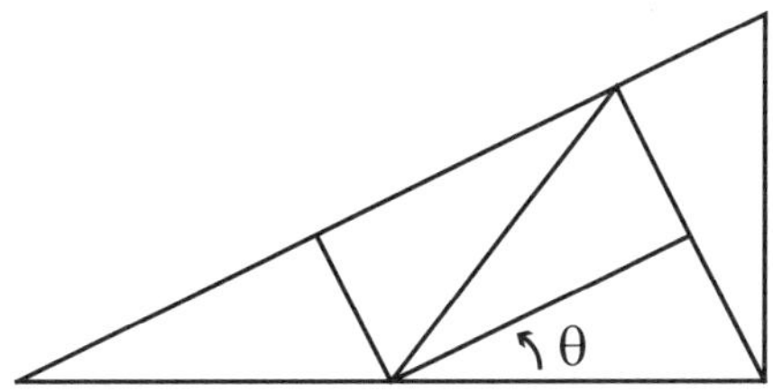

FIGURE 7. The *pinwheel* decomposition: One prototile is irrationally rotated relative to the others.

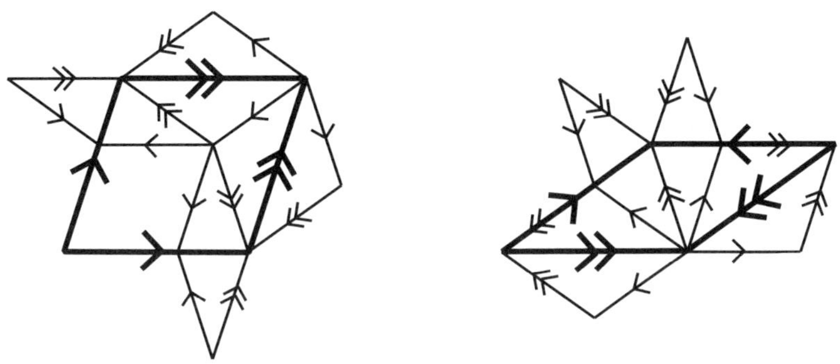

FIGURE 8. The imperfect decomposition of the rhombic Penrose tilings.

EXAMPLE 4.5. (The binary tiling decomposition). The decomposition shown in Figure 9 is from [**GL92**]. A patch of a tiling x in the corresponding substitution tiling space is shown in Figure 10. A perfect version of this decomposition is shown in Figure 13.

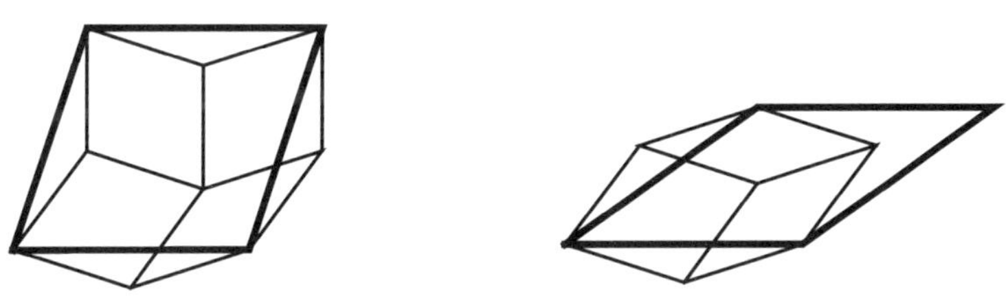

FIGURE 9. The imperfect "binary" tiling decomposition C_b is self-similar. In the corresponding tiling substitution $S_b = \lambda M C_b$ the similarity $L = \lambda M$ includes a rotation M by $2\pi/40$. Note, however, that $C_b^2(\mathcal{T}) \subseteq L^{-2}\mathcal{T}$, so there is a version of S_b^2 with no rotation.

4.2. Properties of tiling substitutions.

DEFINITION 4.6. Let $S = LC$ be a tiling substitution on $\mathcal{T}$ where $X_{\mathcal{T}}$ has finite local complexity. We say S satisfies the 2-*patch closure property* if $S(\mathcal{T}^{(2)}) \subseteq \mathcal{T}^*$.

From now on, unless we say otherwise, we will assume every tiling substitution S satisfies the 2-patch closure property. The reason[6] for this assumption is that a

[6]The need for such an assumption was pointed out to the author by Natalie Priebe Frank.

FIGURE 10. A patch of binary tiling.

tiling substitution $S : \mathcal{T} \to \mathcal{T}^*$ (satisfying 2-patch closure) can be extended to a mapping $S : X_{\mathcal{T}} \to X_{\mathcal{T}}$. This mapping performs the decomposition C, viewed as a mapping $C : X_{\mathcal{T}} \to X_{L^{-1}\mathcal{T}}$, and then applies the linear expansion L to the entire tiling. Note that C satisfies *translation invariance*

$$T^{\mathbf{t}}C = CT^{\mathbf{t}} \text{ for all } \mathbf{t} \in \mathbb{R}^d. \tag{4.2}$$

EXERCISE 4. Show that both C and S are continuous in the tiling topology and that S satisfies $T^{L\mathbf{t}}S = ST^{\mathbf{t}}$.

We call a tiling substitution $S = LC$ *invertible* if it is 1:1. In this case S has a continuous inverse $S^{-1} = L^{-1}C^{-1}$ on $S(X_{\mathcal{T}})$. Then C^{-1} is a continuous and translation invariant mapping, which is called a *composition.*

Let us denote $\mathcal{T} = \{D_1, \ldots, D_n\}$. We define the *structure matrix* A of the tiling substitution $S = LC$ to be $n \times n$ matrix with entries $A_{i,j}$ equal to the number of prototiles equivalent to D_i that occur in $S(D_j)$. A tiling substitution is called *primitive* if $A^k > 0$ for some $k > 0$.

4.3. Substitution tiling spaces. Let S be a tiling substitution on $\mathcal{T}$. Take $D \in \mathcal{T}$ and define a sequence of patches x_k inductively: $x_1 = \{D\}$, and $x_k = S(x_{k-1})$, $k > 1$. Since we assume S satisfies the 2-patch closure property, it follows that $x_k \in \mathcal{T}^*$ for all k.

The patches x_k are used to define a tiling space as follows. First we define a set $\mathcal{F}_S$ of forbidden patches by stipulating that $y \in \mathcal{T}^*$ is forbidden if it is not a sub-patch of x_k for any k. The tiling space $X_S = X_{\mathcal{F}_S} \subseteq X_{\mathcal{T}}$ is called the *substitution tiling space* corresponding to S. The dynamical system (X_S, T) is called a *substitution tiling dynamical system.*

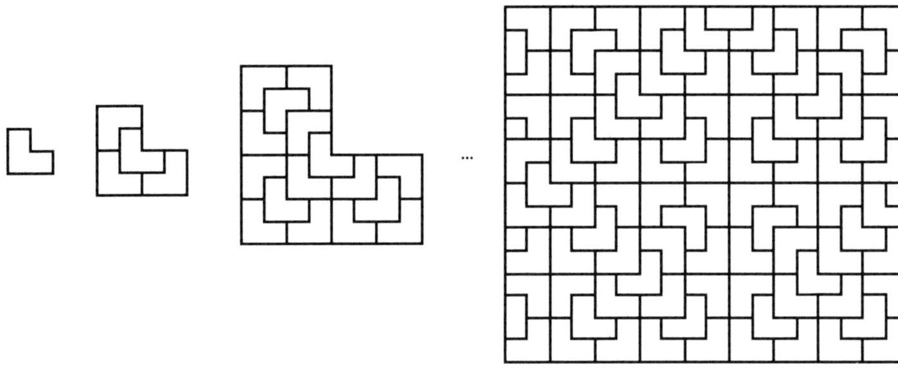

FIGURE 11. The patches x_k, $k = 1, 2, 3$, for the chair tilingsubstitution (Figure 4(a)), and a patch of *chair tiling.*

LEMMA 4.7. *Let $X_S \subseteq X_{\mathcal{T}}$ be the substitution tiling space corresponding to a primitive tiling substitution S on $\mathcal{T}$. Then $X_S \neq \emptyset$ and X_S is independent of the initial tile $D \in \mathcal{T}$. Moreover $S(X_S) \subseteq X_S$.*

PROOF. First we observe that $x_k \in \mathcal{T}^+$ for all k, since by the definition of $\mathcal{F}_S$, no patch x_k contains any forbidden sub-patches. We also note that since L is expansive, $\mathrm{diam}(\mathrm{supp}(x_k)) \to \infty$. Thus given $r > 0$, there exists $\mathbf{t} \in \mathbb{R}^d$ so that $B_r = \mathbf{t} \subseteq \mathrm{supp}(x_k)$ for k sufficiently large. It follows from the Extension Theorem that $X_S = X_{\mathcal{F}_S} \neq \emptyset$. The independence of X_S from the choice of D follows from primitivity. The S-invariance is clear. □

COROLLARY 4.8. *Let $\mathcal{P}$ be the set of marked rhombic Penrose tiles shown in Figure* 2, *and let $\mathcal{P}_1$ be the marked triangular Penrose tiles shown in Figure* 6. *Then the corresponding finite type tiling spaces $X_{\mathcal{P}}$ and $X_{\mathcal{P}_1}$ are nonempty, i.e.,* Penrose tilings exist.

PROOF. For the two tiling substitutions, S from Figure 8 and the substitution S_1 from Figure 6, the corresponding substitution tiling spaces satisfy $X_{\mathcal{P}} \supseteq X_S \neq \emptyset$ and $X_{\mathcal{P}_1} \supseteq X_{S_1} \neq \emptyset$. □

This is essentially Penrose's argument, although he did not couch it in dynamical terms. We conclude with a result that characterizes exactly which tilings belong to a substitution tiling space.

PROPOSITION 4.9. *Let $X_S \subseteq X_{\mathcal{T}}$ be a substitution tiling space for a primitive tiling substitution S. Let $x \in X_{\mathcal{T}}$. Then $x \in X_S$ if and only if there is an infinite sequence $x_n \in X_S$ so that $S^n x_n = x$ (i.e., x has infinitely many S-preimages).*

EXERCISE 5. Prove Proposition 4.9.

COROLLARY 4.10. *For Penrose tilings, the finite type tiling spaces are the same as the substitution tiling spaces: $X_{\mathcal{P}} = X_S$ and $X_{\mathcal{P}_1} = X_{S_1}$.*

PROOF. By Proposition 4.9 and Corollary 4.8 it suffices to show S^{-1} and S_1^{-1} exist on $X_{\mathcal{P}}$ and $X_{\mathcal{P}_1}$ respectively. This is easy to see by inspection for $X_{\mathcal{P}_1}$. Although it is a little harder to see for $X_{\mathcal{P}}$, it is also true in that case.

Alternatively, this result for $X_{\mathcal{P}}$ also follows from $X_{\mathcal{P}_1}$ using Example 5.20 below. □

5. Applications of topological dynamics

5.1. Repetitivity and minimality. Tilings x constructed from aperiodic sets $\mathcal{Q}$ of prototiles, as well as substitution tilings, tend to have the following curious property.

DEFINITION 5.1. A tiling x is called *repetitive* if for any patch y in x there is an $r > 0$ such that for any $\mathbf{t} \in \mathbb{R}^d$ there is a translation $T^{\mathbf{s}}y$ of y in x such that $\mathrm{supp}(T^{\mathbf{s}}y) \subseteq B_r + \mathbf{t}$.

In other words, a copy of y occurs "nearby" any given location $\mathbf{t}$ in x. Since all periodic tilings are repetitive, we think of repetitivity as a generalization of periodicity.

Let (X, T) be a dynamical system. Let $U \subseteq X$ be open and let $x \in X$. Define the *return set* of x to U to be

$$R(x, U) = \{\mathbf{t} \in \mathbb{R}^d : T^{\mathbf{t}}x \in U\}. \tag{5.1}$$

A set $R \subseteq \mathbb{R}^d$ is called *relatively dense* if there is an $r > 0$ such that every r-ball in $\mathbb{R}^d$ intersects R.

DEFINITION 5.2. A point $x \in X$ is *almost periodic* if $R(x, U)$ is relatively dense for every open $U \subseteq X$ with $R(x, U) \neq \emptyset$.

For a tiling space $X \subseteq X_{\mathcal{T}}$, let $y \in \mathcal{T}^*$ and let $R = \mathrm{supp}(y)$. Define $X(y) = \{x \in X : x[R] = y\}$. For $\epsilon > 0$ define the *cylinder set*

$$U_{y,\epsilon} = T^{B_\epsilon}X(y) = \{T^{\mathbf{t}}x : x \in X(y), \mathbf{t} \in B_\epsilon\}.$$

Clearly $U_{y,\epsilon}$ is open.

Without loss of generality, we can assume by translating that the support of each patch $y \in \mathcal{T}^*$ contains the largest possible ball B_r around the origin. It follows that the cylinder sets $\{U_{y,\epsilon_n} : y \in \mathcal{T}^*, \epsilon_n \to 0\}$ form a basis for the tiling topology on X. In Definition 5.2, it suffices to check only the sets U belonging to this basis. Thus we have the following.

PROPOSITION 5.3. *Let (X, T) be a tiling dynamical system. Then a tiling $x \in X$ is repetitive if and only if it is an almost periodic point*[7].

A dynamical system (X, T) is called *minimal* if there are no proper closed T-invariant subsets of X. For a point $x \in X$ we define its *orbit* by $O(x) = \{T^{\mathbf{t}}x : \mathbf{t} \in \mathbb{R}^d\}$, and its orbit closure $\overline{O(x)} \subseteq X$ to be the closure of $O(x)$ in X.

GOTTSCHALK'S THEOREM. ([**Got44**]) *A dynamical system $(\overline{O(x)}, T)$ is minimal if and only if x is almost periodic.*

It follows from the minimality of the dynamical system $(\overline{O(x)}, T)$ that $\overline{O(y)} = \overline{O(x)}$ for all $y \in \overline{O(x)}$. In this case it follows from Gottschalk's Theorem that y is almost periodic too.

DEFINITION 5.4. Two repetitive tilings x, y are said to be *locally isomorphic* if $\overline{O(x)} = \overline{O(y)}$.

[7]Because of this, the author previously used the term "almost periodic tiling" to mean a repetitive tiling.

Geometrically, two locally isomorphic tilings x and y have exactly the same patches. Dynamically, local isomorphism means x and y belong to the same minimal tiling dynamical system.

A dynamical system (X, T) is *transitive* if there is a single orbit: $O(x) = X$ for all $x \in X$. This is a special case of minimality. A tiling x is periodic if and only if $O(x) = \overline{O(x)}$. In this case $(T, O(x))$ is transitive, and $O(x) = \mathbb{R}^d/\Gamma_x$ is a d-dimensional torus (e.g., $X_{\mathcal{S}} = \mathbb{Z}^d/\mathbb{R}^d$).

We say a tiling is *properly repetitive* if it is repetitive but not periodic. An easy application of Zorn's lemma shows that every dynamical system (X, T) has a minimal T-invariant subset $Y \subseteq X$. It follows that every tiling space contains a repetitive tiling (this argument appears in[**RW92**]). Of course in general this tiling may be periodic, but if we know X is an aperiodic tiling space, then it must contain a properly repetitive tiling.

EXERCISE 6. Starting with the Penrose tiles $\mathcal{P}$, construct a new example of an aperiodic prototile set $\mathcal{P}'$ such that not every tiling $x \in X_{\mathcal{P}'}$ is repetitive.

A dynamical system (X, T) is called *topologically transitive* if there exists $x \in X$ such that $\overline{O(x)} = X$. Clearly transitive implies minimal which implies topologically transitive. In each case the converse is false. Exercise 7 shows that $(X_{\mathcal{S}_n}, T)$ is topologically transitive but not minimal. Later we will show that the Penrose tiling dynamical system $(X_{\mathcal{P}}, T)$ is minimal but not transitive. A minimal dynamical system which is not transitive is called *properly minimal.*

EXERCISE 7. Show that the tiling dynamical system $(X_{\mathcal{S}_n}, T)$ is topologically transitive but not minimal. What can you say about $(X_{\mathcal{R}_n}, T)$? (See Examples 2.5 and 2.6.)

REMARK 5.5. When x periodic, $\overline{O(x)}$ is a torus, i.e., a connected manifold. It turns out that this is an exceptional situation. One can show that if (X, T) is aperiodic and topologically transitive, then for each patch y, $X(y)$ is homeomorphic to a Cantor set. Since for ϵ sufficiently small, $U_{y,\epsilon} = T^{B_\epsilon} X(y)$ is homeomorphic to $B_\epsilon \times X(y)$, every point $x \in X$ has a neighborhood homeomorphic to a product of $\mathbb{R}^d$ and a Cantor set. Such a space is called a *lamination.* In particular, a tiling space X is almost never connected.

5.2. The repetitivity of substitution tilings.

PROPOSITION 5.6. *Let X_S be the substitution tiling space corresponding to an invertible primitive tiling substitution S. Then any X_S is an aperiodic tiling space.*

PROOF. Suppose $T^{\mathbf{t}_0} x = x$ for some $\mathbf{t}_0 \neq 0$. Since C is invertible, (4.2) implies

$$T^{\mathbf{t}_0} C^{-n} x = C^{-n} x \tag{5.2}$$

for all $n \geq 1$. Choose n so large that $T^{\mathbf{t}_0} \mathrm{int}(L^n D) \cap \mathrm{int}(L^n D) \neq \emptyset$ for all $D \in \mathcal{T}$. Since for some $D \in \mathcal{T}$, $L^n D \in x$, this contradicts (5.2). □

REMARK 5.7. In Proposition 5.6, invertibility is also necessary [**Sol98**].

The following generalizes a well known result for discrete substitution dynamical systems (see [**Que87**]).

THEOREM 5.8. *Let X_S be a substitution tiling space corresponding to an primitive tiling substitution S. Then any $x \in X_S$ is repetitive. Moreover, any $x, y \in X_S$ are locally isomorphic. In particular, (X_S, T) is minimal.*

COROLLARY 5.9. *The tiling space X_S corresponding to invertible primitive tiling substitution S consists of properly repetitive tilings.*

PROOF. This follows from Gottschalk's Theorem, Theorem 5.6 and Theorem 5.8. □

PROOF OF THEOREM 5.8. Assume without loss of generality $A > 0$ (otherwise replace S with S^k). Let $x \in X$ and let y be a patch in x. Fix $D_0 \in \mathcal{T}$ and choose k so large that y is a patch in $S^{k-1}(\{D_0\})$. Then y is a patch in $S^k(\{D\})$ for all $D \in \mathcal{T}$.

Since $x \in X$ it follows from Proposition 4.9 that there exists $x_k \in X$ so that $S^k x_k = x$. Let $x'_k = L^k x_k \in X_{L^k\mathcal{T}}$. Note that $C^k x'_k = x$.

Let s be the largest diameter of $L^k D \in L^k\mathcal{T}$ and let $r = 2s$. It follows from the triangle inequality that for any $D' \sim L^k D$, if $\mathbf{t} \in D'$ then $D' \subseteq B_r + \mathbf{t}$.

Thus any r-ball $B_r + \mathbf{t}$ in $\mathbb{R}^d$ contains a tile $D' \in x'_k$, and the patch $C^k(\{D'\})$ in x, which has support D', contains a sub-patch that is a copy of y. □

5.3. Self-affine tilings. Let $S = LC$ be a tiling substitution on $\mathcal{T}$. A tiling $x_0 \in X_{\mathcal{T}}$ is called a *self-affine tiling* with *expansion map* L if $Sx_0 = x_0$. When L is a similarity, x_0 is called a *self-similar tiling*. Self-affine tilings play the same role in the theory of substitution tiling dynamical systems that fixed points play in the theory of discrete substitutions. In particular, self-affine tilings provide a alternative definition for substitution tiling spaces.

THEOREM 5.10. *If S is a primitive tiling substitution then there exists $k > 0$ and $x_0 \in X_S$ such that $S^k x_0 = x_0$.*

PROOF. First, we assume without loss of generality that $A > 0$, since otherwise we can replace S with S^k so that $A^k > 0$. Fixing $D_1 \in \mathcal{T}$, there is a translation of $L^{-1}(D_1)$ in $C(D_1)$. By taking additional powers of S, if necessary, we can assure that $L^{-1}(D_1) \subseteq \operatorname{int}(\operatorname{supp}(D))$. Similarly, there is a sequence D_k of tiles equivalent to D_1 so that for all $k > 0$, $L^{-1}(D_k) \in C(D_{k-1})$ and $L^{-1}(D_k) \subseteq \operatorname{int}(\operatorname{supp}(D_{k-1})) \subseteq \operatorname{int}(D_1)$. Since L is expanding, there exists a unique $\mathbf{c} \in \operatorname{int}(D_1)$ satisfying

$$\mathbf{c} \in \bigcup_{k=1}^{\infty} L^{-k} D_k.$$

The point $\mathbf{c}$ is called a *control point*.

Let $D_0 = D_1 - \mathbf{c}$. This is a tile with a control point at the origin. Define the sequence $x_k = S^k(\{D_0\})$ and use this sequence to construct the substitution tiling space X_S. Because of the location of the control point, we have that x_{k-1} is a sub-patch of x_k for all k, and $x_k = Sx_{k-1}$.

Let $x_0 = \cup_{k\geq 1} x_k$, and note that x_0 is a tiling of $\mathbb{R}^d$ by the choice of an *interior* control point. By Proposition 4.9 we have $x_0 \in X_S$, and also $Sx_0 = x_0$. □

It follows from Gottschalk's Theorem and Theorem 5.8 that for any tiling substitution S, $\overline{O(x)} = X_S$ for any $x \in X_S$. By Theorem 5.10 it follows that there exists a self-affine tiling $x_0 \in X_S$, (i.e., $S^k x_0 = x_0$). Thus for any tiling substitution S there exists a self-affine tiling x_0 such that $X_S = \overline{O(x_0)}$. This hints at the alternative definition of X_S mentioned above. However, as the next example shows it is not completely straightforward.

EXAMPLE 5.11. Let S be the *table* substitution and let y_1 be the patch consisting of two rows of two horizontal table tiles, arranged in a 4×2 rectangle. Note that $y_1 \in \mathcal{F}$, the set of forbidden patches for table tilings. Put $y_n = S^{2(n-1)} y_1$, so that $y_n \subseteq y_{n+1}$. Then $y_0 = \cup_{k \geq 1} y_k$ satisfies $S^2 y_0 = y_0$. However $y_0 \notin X_S$ since it contains the patch $y_1 \subseteq y_0$ and $y_1 \in \mathcal{F}$.

One can show that the patch y_1 in the occurs only at the origin in y_0, but nowhere else. In particular, the self-similar tiling y_0 is not repetitive.

The following result shows the correct way to define a substitution tiling space in terms of self-affine tilings.

PROPOSITION 5.12. *Let $S = LC$ be a primitive tiling substitution and let $x_0 \in X_{\mathcal{T}}$ be repetitive and satisfy $S^k x_0 = x_0$ for some $k \geq 1$. Then $X_S = \overline{O(x_0)}$.*

REMARK 5.13. Suppose we start with a tiling substitution S that we do not assume *a priori* satisfies the 2-patch closure property, but that satisfies $S^k x_0 = x_0$ for some $x_0 \in X_{\mathcal{T}}$. It then follows that S does satisfy 2-patch closure. Once this is known, we can use Proposition 5.12 to define X_S provided we can verify that x_0 is repetitive. This may be easier in practice than verifying that S satisfies 2-patch closure.

Next, we consider the question of what linear maps L can occur as the expansion map for a self-affine tiling.

Let $S = \lambda C$ be a primitive tiling substitution for $d = 1$, and let $Sx_0 = x_0$ be a self-similar tiling of $\mathbb{R}$. If A is the structure matrix for S, then it is easy to see that λ must be the *Perron-Frobenius eigenvalue* of A: the unique real eigenvalue of largest modulus (see Section 6.3 below).

EXAMPLE 5.14. A *discrete substitution* σ is a mapping from a finite alphabet $\mathcal{A} = \{1, \ldots, n\}$ to the set of all non-trivial finite words in the alphabet. For a concrete example, see (8.2) in Section 8.6 below (see also [**Que87**]).

We define the *structure matrix* A of σ to be the $n \times n$ matrix such that the entry $A_{i,j}$ is the number of times the letter i occurs in $\sigma(j)$. Assuming A is primitive, we let $\lambda > 0$ and $\mathbf{a} = (a_1, \ldots, a_n) > 0$ be its Perron-Frobenius eigenvalue and eigenvector. Given a primitive non-negative integer matrix A, it is easy to manufacture a substitution σ with structure matrix A (i.e., this amounts to choosing orders for letters in the words corresponding the columns of A).

Now let $\mathcal{T} = \{[0, a_i] : i = 1, \ldots, n\}$ be a set of prototiles in $\mathbb{R}$ and for convenience, identify $\mathcal{T}$ with $\mathcal{A}$ by identifying $[0, a_i]$ with i. Define a tiling substitution S in such a way that $S([0, a_j])$ is the partition of $\lambda \cdot [0, a_j]$ into translates of the intervals corresponding to $i_1, i_2, \ldots, i_{m_j}$, in order, where $\sigma(j) = i_1 i_2 \ldots i_{m_j}$. It follows that S is a tiling substitution for $d = 1$ with expansion λ. Thus we can construct the corresponding substitution tiling dynamical system (X_S, T), and by Theorem 5.10 there exists a self-similar tiling $x_0 \in X_S$ with $S^k x_0 = x_0$ for some k.

Those numbers λ that can be obtained as Perron-Frobenius eigenvalues of a primitive non-negative integer matrix where classified by Doug Lind [**Lin84**], who called them *Perron numbers*. They consist of all positive real algebraic integers λ such that any Galois conjugate λ' of λ satisfies $|\lambda'| < \lambda$. Modulo some technicalities, we have essentially proved the following observation of Thurston.

PROPOSITION 5.15. (Thurston, [**Thu89**]) *A positive real number* λ *is the expansion for a self-similar tiling of* $\mathbb{R}$ (*or equivalently a* 1*-dimensional tiling substitution*) *if and only if it is a Perron number.*

A similar result holds for $d = 2$ in the the self-similar case.

THEOREM 5.16. (Thurston [**Thu89**], Kenyon [**Ken96**]) *Given* $\lambda \in \mathbb{C}$ *there is a primitive* 2*-dimensional self-similar tiling substitution* $S = \lambda C$ *and a self similar-tiling* $x_0 \in X_S$ *with expansion* λ *if and only if* λ *is a* complex Perron number*: an algebraic integer* λ *such that any Galois conjugate* λ' *of* λ, *except possibly the complex conjugate* $\overline{\lambda}$, *satisfies* $|\lambda'| < |\lambda|$.

We call an expansion $L \in Gl(d, \mathbb{R})$ a *Perron expansion* if its eigenvalues $\Lambda = \{\lambda_1, \ldots, \lambda_d\}$, written with multiplicity, satisfy the condition that for every $\lambda \in \Lambda$ with multiplicity k, if λ' is a Galois conjugate of λ with $|\lambda'| > |\lambda|$, then $\lambda' \in \Lambda$ with multiplicity $k' \geq k$. This idea generalizes both real and complex Perron numbers viewed as expansions of $\mathbb{R}$ and $\mathbb{C} \cong \mathbb{R}^2$ respectively.

THEOREM 5.17. (Kenyon, [**Ken90**]) *If a diagonalizable linear map* $L \in Gl(d, \mathbb{R})$ *is the expansion for a primitive self-similar tiling substitution* $S = LC$, *and* $Sx_0 = x_0$ *for some* $x_0 \in X_S$, *then* L *is a Perron expansion.*

Kenyon [**Ken90**] claims that the converse is also true.

5.4. Local mappings. Local mappings play much the same role in tiling dynamical systems that sliding block codes play in symbolic dynamics. The following version of the definition comes from [**PS01**].

DEFINITION 5.18. A continuous mapping between tiling spaces $Q : X \to Y$ is called a *local mapping* if there is an $r > 0$ so that for all $x \in X$, $Q(x)[\{0\}]$ depends only on $x[B_r]$. We say $Q(x)$ is *locally derivable* from x. If Q is invertible, we say x and $Q(x)$ are *mutually locally derivable.*

EXERCISE 8. Show that a local mapping is continuous and T-equivariant (i.e., $T^{\mathbf{t}}Qx = QT^{\mathbf{t}}x$). Moreover, if a local mapping is invertible then its inverse is also a local mapping. Thus a composition mapping is local.

Now consider two dynamical systems (X, T) and (Y, T). A surjective continuous mapping $Q : X \to Y$ so that $T^{\mathbf{t}}Q = QT^{\mathbf{t}}$ for all $\mathbf{t} \in \mathbb{R}^d$ is called a *factor* mapping, and (Y, T) is called a factor of (X, T). An invertible factor mapping Q is called a *topological conjugacy.* In this case the two dynamical systems are said to be *topologically conjugate.*

LEMMA 5.19. *If* $Q : X \to Y$ *is a surjective local mapping between tiling spaces then it is a factor mapping between tiling dynamical systems. If* Q *is invertible then it is a topological conjugacy.*

EXAMPLE 5.20. (Equivalence of different Penrose tilings). Let $X_{\mathcal{P}}$ denote rhombic Penrose tilings and let let $X_{\mathcal{P}_1}$ be the triangular Penrose tilings. Figure 12 shows how these two types of tilings are mutually locally derivable. Similar invertible local mappings connect these examples to several other famous types of Penrose tilings not discussed here (e.g., the "kites and darts" tilings and Penrose's original "pentagon" tilings: see [**GS87**]). It follows that all the various corresponding Penrose tiling dynamical systems are topologically conjugate.

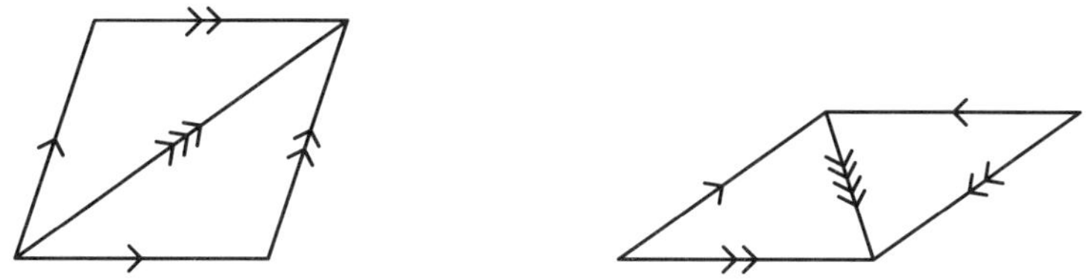

FIGURE 12. The local equivalence between rhombic Penrose tilings and triangular Penrose tilings.

DEFINITION 5.21. A factor mapping $Q : X \to Y$ is *almost* 1:1 if there is a point $y_0 \in Y$ so that $\text{card}Q^{-1}(\{y_0\}) = 1$.

The name of this kind of factor comes from the fact that if (Y, T) is topologically transitive, then $\{y \in Y : \text{card}(Q^{-1}(y)) = 1\}$ is a dense G_δ set.

Now suppose $\mathcal{T}_\#$ is a marked version of a prototile set $\mathcal{T}$ (e.g., in the way $\mathcal{P}$ is a marked version of $\mathcal{R}_5$). Then the local mapping F that erases or "forgets" the markings is an example of a local mapping.

THEOREM 5.22. (Goodman-Strauss [**G-S98**]) *Suppose (X_S, T) is a self-similar substitution tiling dynamical system with $X_S \subseteq X_\mathcal{T}$, for 2-dimensional set $\mathcal{T}$ of prototiles with finite local complexity. Then there exists a marking $\mathcal{T}_\#$ of $\mathcal{T}$, and a local matching rule $\mathcal{T}_\#^{(2)}$ such that the forgetful mapping $F : X_{\mathcal{T}_\#} \to X_\mathcal{T}$ satisfies $F(X_{\mathcal{T}_\#}) = X_S$, and $F : X_{\mathcal{T}_\#} \to X_S$ is almost 1:1.*

In most cases this mapping is *not* invertible (i.e., is not a topological conjugacy). In the terminology of symbolic dynamics, a non finite type factor of a finite type tiling dynamical system should be called *strictly sofic*. It follows that a substitution tiling space is always sofic and usually strictly sofic. An exception is the following. Let $F : \mathcal{P} \to \mathcal{R}_5$ be the mapping that erases the arrows defining the Penrose matching rules. The tilings $X = F(X_\mathcal{P}) \subseteq X_{\mathcal{R}_5}$ are called the *unmarked Penrose tilings*.

THEOREM 5.23. (de Bruijn [**dB81**]) *The marked Penrose tilings are mutually locally derivable with the unmarked Penrose tilings.*

It is interesting to compare these examples to the the situation for 1-dimensional symbolic dynamics. In that case, the intersection between substitution dynamical systems and sofic shifts (or shifts of finite type) consists only of purely periodic examples.

REMARK 5.24. Even though the finite type property is not closed under factorization, it is closed under topological conjugacy. See [**RS01**] for a proof.

To illustrate one further application of local mappings, we return to the idea of an imperfect tiling substitution.

DEFINITION 5.25. An *imperfect tiling substitution* is a mapping $S = LC : X_\mathcal{T} \to X_\mathcal{T}$ where $L \in Gl(d, \mathbb{Z})$ is expansive and $C : X_\mathcal{T} \to X_{L^{-1}\mathcal{T}}$ is a local mapping. For emphasis, the corresponding tiling dynamical system $X_S \subseteq X_\mathcal{T}$ is called an *imperfect substitution* tiling dynamical system.

Note that Examples 4.4 and 4.5 both satisfy this. An imperfect substitution is invertible if C is an invertible local mapping and it is also possible extend the notion of primitivity to this case.

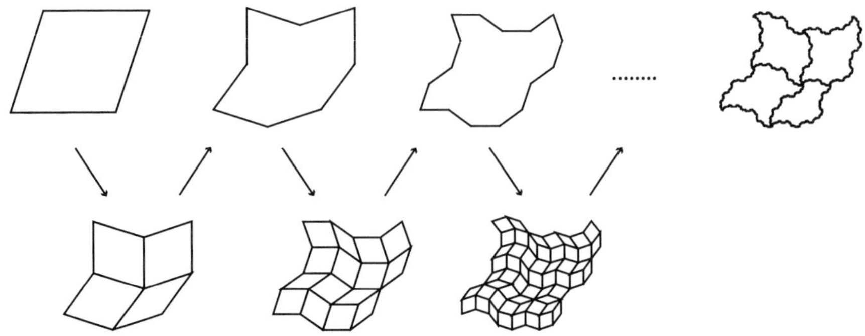

FIGURE 13. Iterating the boundary in the binary tiling decomposition to obtain a perfect decomposition with a fractal boundary.

THEOREM 5.26. (Priebe and Solomyak [**PS01**]) *Suppose $X_S \subseteq X_T$ is a imperfect substitution tiling dynamical system with $d = 2$. Then there is new prototile set T' and a perfect substitution S' on T' so that the substitution tiling dynamical system $X_{S'} \subseteq X_{T'}$ and X_S are topologically conjugate via a local mapping (i.e., corresponding tilings are mutually locally derivable).*

This theorem is proved using a construction called "iterating the boundary" which often results in producing fractal tiles. In Figure 13 this idea is illustrated in the case of the binary tiling system of Example 4.5

REMARK 5.27. One way in which tiling dynamical systems differ from discrete symbolic dynamical systems is the following. The Curtis-Lyndon-Hedlund Theorem [**Hed69**] says that any factor mapping between discrete symbolic dynamical systems is implemented by a sliding block code. In tiling dynamical systems the equivalent question is whether every topological conjugacy is implemented by a local mapping. i.e., is the converse to Lemma 5.19 true? A negative answer was provided by Petersen [**Pet99**] and Radin and Sadun [**RS01**].

5.5. Incongruent tilings. In the case that x is a periodic tiling one has $O(x) = \overline{O(x)}$. In other words there is a single orbit. It follows that, up to translation or congruence, there is just a single tiling.

THEOREM 5.28. *If x is a properly repetitive tiling then the number of orbits in $\overline{O(x)}$ is uncountable. There are uncountably many incongruent tilings in a local isomorphism class.*

This follows directly form the next lemma, which illustrates the power of simple topological ideas in this subject.

PROPOSITION 5.29. *Suppose (X, T) is a minimal dynamical system with T an action of $\mathbb{R}^d$. Let $\Omega \subseteq X$ be such that X can be expressed as a disjoint union of orbits $X = \cup_{x \in \Omega} O(x)$. Then either* $\mathrm{card}(\Omega) = 1$ *or* $\mathrm{card}(\Omega) > \aleph_0$.

PROOF. If Ω is finite then $\mathrm{card}(\Omega) = 1$ since (X, T) is minimal. Thus we suppose Ω is infinite.

For $x \in \Omega$, write x in terms of its tiles $x = \{D_1, D_2, \ldots\}$. Let $V_i(x) = \{T^{\mathbf{t}}x : 0 \in T^{\mathbf{t}}D_i\}$. Then $O(x) = \cup_{i=1}^{\infty} V_i(x)$ is a countable decomposition of $O(x)$ into nowhere dense sets (i.e., they have an empty interior). Thus $X = \bigcup_{x \in \Omega} \bigcup_{i=1}^{\infty} V_i(x)$.

But since X is a compact metric space, it follows from the Baire Category Theorem (see [**Mun75**]) that X is not a countable union of nowhere dense sets. This implies Ω is uncountable. □

COROLLARY 5.30. (Penrose) *In any Penrose tiling space X there are uncountably many incongruent Penrose tilings.*

5.6. Quasicrystallography. Let $M(d)$ denote the set of all rigid motions of $\mathbb{R}^d$ (i.e., the set of congruence transformations). Let $O(d)$ be the subgroup of $M(d)$ fixing the origin. Denote the subgroup of translations in $M(d)$ by $\mathbb{R}^d$.

Suppose x is a periodic tiling. The following ideas are basic to mathematical crystallography (see [**Sen95**]). The *space group* or *symmetry group* of x is defined $G_x = \{M \in M(d) : Mx = x\}$. The translation group Γ_x is a normal subgroup, and the quotient $H_x = G_x/\Gamma_x$, called the *point group* is isomorphic to a finite subgroup of $O(n)$. The *Crystallographic Restriction* is the theorem which says that in any dimension d, there are only finitely many possibilities for H_x. In particular, for $d = 2$ no $M \in H_x$ can have order 5.

Now we sketch the outlines of a theory of *quasicrystallography* (see [**Rob96b**] for more details). For a tiling $x \in X$ define $G_{x,X} = \{M \in M(d) : Mx \in X\}$.

LEMMA 5.31. [**Rob96b**] *If (X, T) is minimal then $G_{x,X} = G_{y,X}$ for all $x, y \in X$.*

EXERCISE 9. Prove Lemma 5.31.

In the minimal case, we write G_X. It follows that G_X is a closed subgroup of $M(d)$ containing $\mathbb{R}^d$ as a normal subgroup. We call G_X the *quasisymmetry group* and we call the quotient $H_X = G_X/\mathbb{R}^d$ the *quasicrystallographic* point group. The algebraic situation is simpler than in the case of symmetry groups. One always has that G_X is a semi-direct product of $\mathbb{R}^d$ and H_X. In particular, H_X is isomorphic to a closed subgroup of $O(d)$.

PROPOSITION 5.32. *For the Penrose tiling space $X_{\mathcal{P}}$, $H_{X_{\mathcal{P}}} = D_{10}$ (the dihedral group of order 20). For the Pinwheel tiling space X, $H_X = O(2)$. Thus H_X contains the circle $\mathbb{T}$ as a subgroup.*

In fact, one can construct examples X so that H_X contains any finite order rotation. It follows that there is no Crystallographic Restriction for quasicrystals.

6. Applications of ergodic theory

6.1. Measures. Let (X, T) be a dynamical system. In this section we discuss the set $M(X)$ of Borel probability measures in X.

Without going into a lot of details (see for example [**Wal82**]) we mention that a *measure* $\mu \in M(X)$ is a function that assigns a number $0 < \mu(E) < 1$ to a Borel set $E \subset X$. One way to interpret μ is as a "probability law" in which $\mu(E)$ measures the probability that a randomly chosen point $x \in X$ belongs to E. Borel sets can be complicated, but include all open sets, closed sets, and most of the other sets that typically arise in practice. We will assume all sets mentioned are Borel sets.

The integral of a function with respect to a measure μ is denoted $\int_X f(x)\, d\mu$. If we think of f as a random variable on X, then the integral is its expectation.

Of particular interest to us will be T-invariant measures. These are the measures that satisfy $\mu(T^{\mathbf{t}}E) = \mu(E)$ or $f(T^{\mathbf{t}}x)\, d\mu = \int_X f(x)\, d\mu$ for all $\mathbf{t} \in \mathbb{R}^d$. We denote the set of all invariant measures by $M(X,T)$. One can show that always $M(X,T) \neq \emptyset$.

An important feature of measure theory is that a measure μ is completely determined by its values on a collection of sets smaller than the collection of Borel sets. For a tiling dynamical system (X,T), a measure $\mu \in M(X,T)$ is determined by its values on cylinder sets. One can show that there is a function $\mu_0 : \mathcal{T}^* \to \mathbb{R}$ such that

$$\mu(U_{y,\epsilon}) = \mu_0(y)\mathrm{Vol}(B_1)\epsilon^d, \tag{6.1}$$

for ϵ sufficiently small. This generalizes a similar and well known result that holds for discrete symbolic dynamical systems.

6.2. Unique ergodicity. An invariant measure is said to be *ergodic* if $T^{\mathbf{t}}E = E$ for all $\mathbf{t} \in \mathbb{R}^d$ implies $\mu(E) = 0$ or $\mu(E) = 1$ (note the similarity to the idea of minimality). A dynamical system is called *uniquely ergodic* if $M(X,T) = \{\mu\}$. In this case μ is always an ergodic measure.

THEOREM 6.1. *If S is a primitive tiling substitution then the corresponding tiling dynamical system (X_S,T) is uniquely ergodic.*

A similar result is well known in the case of discrete substitution systems (see [**Que87**]). We will prove Theorem 6.1 below. Later we will also discuss a different kind of uniquely ergodic tiling dynamical system, but first we discuss the consequences of unique ergodicity.

Uniquely ergodic dynamical systems satisfy the following especially strong version of the *Ergodic Theorem*.

THEOREM 6.2. *If a dynamical system (X,T) is uniquely ergodic then for all complex valued continuous functions f on X*

$$\lim_{t\to\infty} \frac{1}{\mathrm{Vol}(B_t)} \int_{B_t} f(T^{\mathbf{t}}x)\, d\mathbf{t} = \int_X f(x) d\mu, \tag{6.2}$$

where the expression on the left, viewed as a function of x, converges uniformly *to the integral the right (a constant). Conversely, if for all continuous f and for all x, the limit in* (6.2) *exists, then (X,T) is uniquely ergodic.*

Now let us assume (X,T) is a minimal uniquely ergodic tiling dynamical system. Let $x \in X$ be a tiling and let $y \in \mathcal{T}^*$ be a patch that occurs in x. We know that the occurrences of y are relatively dense, but suppose we want a more quantitative description of this repetitivity. For simplicity we assume, as above, that $\mathrm{supp}(y)$ contains a maximal ball around the origin. Let $P(x,y) = \{\mathbf{t} \in \mathbb{R}^d : T^{\mathbf{t}}y \subseteq x\}$. Note that this is a subset of $R(x, U_{y,\epsilon})$. Recall that the *characteristic function* of a set U is

$$\chi_U(x) = \begin{cases} 1 & \text{if } x \in U, \\ 0 & \text{if } x \notin U. \end{cases}$$

We have in particular $\int_X \chi_U(x) d\,\mu = \mu(U)$. It turns out that even though characteristic functions of cylinder sets are not continuous, Theorem 6.2 still holds for them.

COROLLARY 6.3. *A tiling dynamical system (X,T) is uniquely ergodic if and only if for any $x \in X$ and any $y \in \mathcal{T}^*$ the following limit exists:*

$$\lim_{t\to\infty} \frac{1}{\mathrm{Vol}(B_t)} \mathrm{card}(B_t \cap P(x,y)). \tag{6.3}$$

If (X,T) is uniquely ergodic, then the value of limit (6.3) is $\mu_0(y)$.

This result explains the combinatorial and geometric meaning of the unique invariant measure in the uniquely ergodic case: it determines the frequency of all the different tiling patches y in all the tilings $x \in X$.

In fact, a similar result holds under the weaker assumption that $\mu \in M(X,T)$ is just ergodic. We say x is *generic* for μ if (6.3) holds. That is, every patch has a well defined frequency. It follows from the Birkhoff Ergodic Theorem (see [**Wal82**]) that μ a.e. $x \in X$ is generic. If (X,T) is not uniquely ergodic, different tilings will be generic for different ergodic measures. This will result in different patch frequencies for different tilings. Moreover, by Corollary 6.3 there will always be some tilings $x \in X$ so that the limit (6.3) diverges. In these tilings, certain patches will not have well defined frequencies.

6.3. Perron Frobenius theory. In this section we present the main part of the proof of Theorem 6.2. We use a well known argument (see [**Que87**]) based on the Perron-Frobenius Theorem to show that in a substitution tiling space, every prototile occurs in every tiling with a well defined density. Unfortunately this is not quite enough to prove Theorem 6.2. In the next section we show how to generalize the idea of a higher block code from symbolic dynamics to tiling spaces, and how to use this idea to finish the argument.

THEOREM 6.4 (Perron-Frobenius Theorem (see [**Rue69**])). *Let $A \geq 0$ be a real square matrix with $A^k > 0$ for some $k \geq 1$. Then there is a simple positive eigenvalue $\omega > 0$ with $\omega > |\omega'|$ for all other eigenvalues ω'. Let $\mathbf{a}$ and $\mathbf{b}$ be the eigenvectors corresponding to ω for A and A^T. Then $\mathbf{a}, \mathbf{b} > 0$ and for any $\mathbf{v} \in \mathbb{R}^d$*

$$\lim_{n\to\infty} \frac{1}{\omega^n} A^n \mathbf{v} = (\mathbf{b}\cdot\mathbf{v})\mathbf{a}. \tag{6.4}$$

The eigenvalue $\omega > 0$ and eigenvector $\mathbf{a} > 0$ are called the *Perron-Frobenius eigenvalue* and *Perron-Frobenius eigenvector* of A.

COROLLARY 6.5. *let A be the structure matrix for a primitive tiling substitution $S = LC$ and let ω be the Perron-Frobenius eigenvalue of A. Then $\omega = \det(L)$.*

This is because both ω and $\det(L)$ measure how the substitution S expands volumes.

COROLLARY 6.6. *Let X be a primitive substitution tiling space. For $D_i \in \mathcal{T}$ let $D_i^n = \mathrm{supp}(S^n D_i)$. Then for any $x \in X$ and $D_i, D_j \in \mathcal{T}$ the limit*

$$\lim_{n\to\infty} \frac{1}{\mathrm{Vol}(D_i^n)} \mathrm{card}(D_i^n \cap P(x,\{D_j\}))$$

exists.

PROOF. Since $\omega^n = \det(L^n)$, we have $\mathrm{Vol}(D_i^n) = \omega^n \mathrm{Vol}(D_i)$. Write $d_i = \mathrm{Vol}(D_i)$. Let $A_{i,j}^n$ be the i,jth entry of A^n. Note that $A_{i,j}^n$ is the number of tiles equivalent to D_j that occur in $S^n(\{D_i\})$, so

$$A_{i,j}^n = \mathrm{card}(D_i^n \cap P(x,\{D_j\})). \tag{6.5}$$

Since $A^n_{i,j} = \mathbf{e}_i \cdot (A^n \mathbf{e}_j)$, it follows from (6.4)

$$\begin{aligned}\lim_{n\to\infty} \frac{1}{\mathrm{Vol}(D_i^n)} \mathrm{card}(D_i^n \cap P(x, \{D_j\})) &= d_i^{-1} \lim_{n\to\infty} \frac{1}{\omega^n} A^n_{i,j} \\ &= d_i^{-1} \lim_{n\to\infty} \frac{1}{\omega^n} \mathbf{e}_i \cdot (A^n \mathbf{e}_j) \\ &= d_i^{-1} \mathbf{e}_i \cdot \left(\lim_{n\to\infty} \frac{1}{\omega^n} (A^n \mathbf{e}_j) \right) \\ &= d_i^{-1} (\mathbf{b} \cdot \mathbf{e}_j)(\mathbf{e}_i \cdot \mathbf{a}) \\ &= \frac{a_i b_j}{d_j},\end{aligned}$$

where $\mathbf{a} = (a_1, \ldots, a_d)$ and $\mathbf{b} = (b_1, \ldots, b_d)$. □

With a bit more care one can prove the following refinement of the above.

COROLLARY 6.7. *Let X be a primitive substitution tiling space. Then for any $x \in X$ and $D \in \mathcal{T}$ the limit*

$$\lim_{r\to\infty} \frac{1}{\mathrm{Vol}(B_r)} \mathrm{card}(B_r \cap P(x, \{D\})) \tag{6.6}$$

exists.

6.4. Higher patch tiles. This section discusses a technical result needed to complete the proof of Theorem 6.1.

Let (X, T) be a tiling dynamical system $X \subseteq X_{\mathcal{T}}$. Fix $r > 0$ and let $y_1, y_2, \ldots, y_m$ be the equivalence classes of patches $x[B_r]$, $x \in X$. Let

$$E_i = \overline{\{\mathbf{t} \in \mathbb{R}^d : x[B_r + \mathbf{t}] \sim y_i\}},$$

where we assume E_i has nonempty interior. Define a new tiling x' by subdividing each tile $D \in x$ into the smaller tiles $D \cap E_i$, $i = 1, \ldots, m$. Up to equivalence there are only finitely prototiles in x' for all $x \in X$ and we denote this new prototile set by $\mathcal{T}_r$. Moreover, the mapping $H_r : X_{\mathcal{T}} \to X_{\mathcal{T}_r}$ is clearly an invertible decomposition mapping. It is called *higher patch mapping*. This is similar to the idea of a *higher block code* in symbolic dynamics (see [**LM95**], [**Que87**]).

PROPOSITION 6.8. *If S is a tiling substitution on $X_{\mathcal{T}}$ then $S_r = H_r^{-1} S H_r$ is a tiling substitution on $X_{\mathcal{T}_r}$. If S is primitive then so is S_r.*

Theorem 6.1 now follows by applying Proposition 6.8 and Corollary 6.7.

7. Mixing properties

7.1. Mixing and eigenvalues: the geometric interpretation. Let (X, T) be a dynamical system and $\mu \in M(X, T)$. A complex function $f \in L^2(X, \mu)$ is called an *eigenfunction* if there exists a corresponding *eigenvalue* $\mathbf{w} \in \mathbb{R}^d$ such that

$$f(T^{\mathbf{t}} x) = e^{2\pi i \langle \mathbf{t}, \mathbf{w} \rangle} f(x), \tag{7.1}$$

for μ a.e. x. Note that this "eigenvalue" is actually a vector! In Physics this might be called a "wave vector."

A constant function f is always an eigenfunction, corresponding to $\mathbf{w} = \mathbf{0}$. Ergodicity is equivalent to $\mathbf{w} = \mathbf{0}$ being a simple eigenvalue. Moreover, in the ergodic case all the eigenvalues are simple, and the set Σ of eigenvalues is a countable subgroup of $\mathbb{R}^d$ (see [**Wal82**]).

If the only eigenfunctions are constants, then T is said to be *weakly mixing.* The opposite situation is called *pure discrete spectrum*; it occurs when the eigenfunctions have a dense span in $L^2(X, \mu)$. An eigenfunction f is continuous if it is equal μ a.e. to a continuous function.

A dynamical system is called *strongly mixing* (or just *mixing*)if for any Borel sets A and B

$$\lim_{||\mathbf{t}||\to\infty} \mu(T^{\mathbf{t}}A \cap B) = \mu(A)\mu(B). \tag{7.2}$$

A well known theorem (see [**Wal82**]) says that weak mixing is equivalent to (7.2) holding except on a set of $\mathbf{t}$ of density zero (this set depends on A and B).

Now consider a tiling dynamical system (X, T). As we will discuss below, the eigenvalues are related to the the "diffraction" properties of tilings $x \in X$. Heuristically, such diffraction is caused by constructive reinforcement of waves reflecting off atoms, usually thought of as being located at the vertices of a tiling. When a tiling exhibits diffraction it can be interpreted as evidence that the tiling has some sort of long range spatial order in the arrangement of its tiles. Periodic tilings always diffract, but as we will see below, so do some properly repetitive tilings.

Conversely, if a tiling system satisfies a mixing property (i.e., a lack of diffraction) then it indicates that its tilings enjoy some sort of long-range spatial disorder. Consider, for example, (7.2) applied to a pair of cylinder sets U_{ϵ,y_1} and U_{ϵ,y_2} in a mixing tiling dynamical system (X, T). For a randomly chosen $x \in X$ and for $\mathbf{t}$ sufficiently large, the probability of seeing y_1 and $T^{\mathbf{t}}y_2$, (up to an ϵ-translation), is approximately $\epsilon^2 \mathrm{Vol}(B_1)^2 \mu_0(y_1)\mu_0(y_2)$. Thus, the knowledge that y_1 sits at one place in x is approximately statistically independent of the knowledge that a copy of y_2 sits at any particular distant location.

7.2. Weakly mixing tiling spaces. There are two known mechanisms for producing weakly mixing tiling dynamical systems. The first is related to the algebraic properties of the eigenvalues of the expansion. It generalizes ideas from the theory of discrete 1-dimensional substitutions.

Let $D \in \mathcal{T}$ and $x \in X_{\mathcal{T}}$. Define

$$\Xi(x) = \{\mathbf{t} \in \mathbb{R}^d : \exists D_1, D_2 \in x, D_2 = D_1 - \mathbf{t}\}.$$

If (X, T) is a properly minimal tiling dynamical system, then $\Xi(x)$ is the same for all tilings $x \in X$, and we write $\Xi(X)$. In addition, $\Xi(X)$ satisfies

$$\overline{\{\mathbf{t}/||\mathbf{t}|| : \mathbf{t} \in \Xi(X), \mathbf{t} \neq 0\}} = S^{d-1} \subseteq \mathbb{R}^d \tag{7.3}$$

(see [**Sol97**], Proof of Theorem 4.4). Note that this is the case when $X = X_S$ is a primitive substitution tiling space.

THEOREM 7.1. (Solomyak [**Sol97**]) *A number $\mathbf{w} \in \mathbb{R}^d$ is an eigenvalue for an invertible primitive self-affine substitution tiling system (X_S, T) with $S = LC$ if and only if*

$$\lim_{n\to\infty} e^{2\pi i \langle L^n \mathbf{t}, \mathbf{w}\rangle} = 1 \tag{7.4}$$

for all $\mathbf{t} \in \Xi(X_S)$. Moreover, the eigenfunctions can always be chosen to be continuous.

This is a combination of Theorems 4.3 and 5.1 in [**Sol97**]. It generalizes a similar result of Host [**Hos86**] for 1-dimensional discrete substitutions.

REMARK 7.2. In certain non-weakly mixing cases Solomyak [**Sol97**] describes some eigenvalues explicitly.

Our interest here is on weak mixing, and for this purpose the beauty of Theorem 7.1 is that it reduces the question to number theory. A real algebraic integer $\lambda > 1$ is called a *Pisot number* if all of its Galois conjugates λ' satisfy $|\lambda'| < 1$. *Pisot's Theorem* says that a positive real algebraic number λ satisfying $e^{2\pi i\omega\lambda^n} \to 1$ for some for some $\omega \in \mathbb{R}$ must be a Pisot number (see [**Sal63**]). A *complex Pisot number* is a complex algebraic integer λ all of whose Galois conjugates λ', except its complex conjugate, satisfy $|\lambda'| < 1$. The following generalization of this idea is due to Mauduit [**Mau89**].

DEFINITION 7.3. A set $\Lambda' = \{\lambda_1, \ldots, \lambda_{d'}\}$ of distinct algebraic integers with $|\lambda_i| \geq 1$ is called a *Pisot family* if λ' is Galois conjugate of some $\lambda \in \Lambda'$ with $\lambda' \notin \Lambda'$, then $|\lambda'| < 1$. Otherwise Λ' is called *non-Pisot*.

A real Pisot number by itself is a Pisot family, as is a complex Pisot number together with its complex conjugate. The next result generalizes Pisot's Theorem.

THEOREM 7.4. (Mauduit, [**Mau89**]) *If $\Lambda' = \{\lambda_1, \ldots, \lambda_{d'}\}$ is set of distinct algebraic numbers such that*

$$\lim_{n\to\infty} e^{2\pi i \sum_{i=1}^{d'} v_i \lambda_i^n} = 1$$

for some[8] $(v_1, \ldots, v_{d'}) \in (\mathbb{C}\backslash\{0\})^{d'}$, *then Λ' is a Pisot family.*

A nonempty set Λ' of distinct algebraic integers can be written as a disjoint union

$$\Lambda' = \Lambda_1 \cup \Lambda_2 \cup \cdots \cup \Lambda_\ell \tag{7.5}$$

where for each i there exists a monic irreducible polynomial $p_i \in \mathbb{Z}[t]$ (the minimal polynomial) such that $p_i(\lambda) = 0$ for all $\lambda \in \Lambda_i$.

DEFINITION 7.5. We say Λ' is *totally non-Pisot* if each Λ_i in the decomposition (7.5) is non-Pisot.

It is clear that a totally non-Pisot family Λ' is non-Pisot, and moreover, any nonempty subset $\Lambda'' \subseteq \Lambda'$ is totally non-Pisot. We call an expansion L *totally non-Pisot* if its set Λ' of eigenvalues, written without multiplicity, is a totally non-Pisot family.

THEOREM 7.6. *Suppose $S = LC$ is a primitive invertible tiling substitution such that the expansion L is diagonalizable and totally non-Pisot. Then the substitution tiling system (X_S, T) is weakly mixing.*

This is essentially due to Solomyak [**Sol97**], although our formulation is different.

---

[8] Mauduit [**Mau89**] proves this for $(\mathbb{R}\backslash\{0\})^{d'}$, but the proof works in the complex case.

PROOF. Let $\Lambda' = \{\lambda_1, \ldots, \lambda_{d'}\}$ be the eigenvalues of L, written without multiplicity. Then Λ' is totally non-Pisot. Given $\mathbf{t} \in \mathbb{R}^d$ express $\mathbf{t} = \sum_{j=1}^{d'} P_j \mathbf{t}$ where P_j is the projection to the eigenspace for λ_j, parallel to the all the other eigenspaces, and $\mathbf{t}_j = P_j\,\mathbf{t} \in \mathbb{C}^d$. Then $L^n\mathbf{t} = \sum_{j=1}^{d'} \lambda_j^n \mathbf{t}_j$.

Now suppose $\mathbf{w} \neq \mathbf{0}$ is an eigenvalue for (X_S, T). Then by Theorem 7.1 we have

$$(7.6) \qquad 1 = \lim_{n\to\infty} e^{2\pi i \langle L^n \mathbf{t}, \mathbf{w}\rangle} = \lim_{n\to\infty} e^{2\pi i \sum_{j=1}^{d'} \lambda_j^n \langle \mathbf{t}_j, \mathbf{w}\rangle} = \lim_{n\to\infty} e^{2\pi i \sum_{j=1}^{d'} \lambda_j^n \langle \mathbf{t}, P_j^* \mathbf{w}\rangle},$$

where $\langle \mathbf{t}, \mathbf{w}\rangle = \mathbf{t}\cdot\overline{\mathbf{w}}$ is the inner product on $\mathbb{C}^d$, and P_j^* is the complex-conjugate transpose of P_j.

Let $\Lambda''(\mathbf{w}) = \{\lambda_j \in \Lambda' : P_j^*(\mathbf{w}) \neq 0\}$, and note that $\Lambda''(\mathbf{w}) \neq \emptyset$ since L is diagonalizable. By (7.3) there exists $\mathbf{t} \in \Xi(X)$ such that $\langle \mathbf{t}, P_j^*\mathbf{w}\rangle \neq 0$ for all j such that $\lambda_j \in \Lambda''(\mathbf{w})$. Applying (7.6), it follows from Theorem 7.4 that $\Lambda''(\mathbf{w})$ is a Pisot family. But since $\Lambda''(\mathbf{w}) \subseteq \Lambda'$, this contradicts the fact that Λ' is totally non-Pisot □

REMARK 7.7.

(1) Suppose $d \geq 1$ and let $S = \lambda C$ be a self-similar tiling substitution with $\lambda \in \mathbb{R}$. If λ is not real Pisot then (X_S, T) is weakly mixing.
(2) Suppose $d = 2$ and $S = \lambda C$, where $\lambda \in \mathbb{C}\backslash\mathbb{R}$. If λ is not complex Pisot, then (X_S, T) is weakly mixing.
(3) In both cases above, for $d = 2$, the converse is also true, [**Sol97**].

COROLLARY 7.8. *The binary tiling dynamical system (X_{S_b}, T) is weakly mixing.*

PROOF. We use the fact that $C_b^2(\mathcal{T}) \subseteq \lambda^{-2}\mathcal{T}^*$, $\lambda \in \mathbb{R}$, so that $S_b^2 = \lambda C_b^2$. The structure matrix for S_b^2 is A^2, where $A = \begin{pmatrix} 3 & 1 \\ 1 & 2 \end{pmatrix}$. Thus, the expansion λ for S_b^2 is the Perron-Frobenius eigenvalue λ for A, which is not real Pisot. □

The reader should compare the disordered appearance of binary tilings to the more regular appearance of the Penrose tilings. As we will see below, Penrose tiling dynamical systems have pure discrete spectrum.

REMARK 7.9. Let $S = \lambda C$ be a tiling substitution with $d = 2$ and $\lambda \in \mathbb{R}\backslash\mathbb{C}$. We show here that in order to establish that (X_S, T) is weakly mixing it is *not* sufficient (i) that $|\lambda|$ is not real Pisot, nor (ii) that $|\lambda|^2 = \omega$ (where ω is the Perron-Frobenius eigenvector for A) is not real Pisot.

For (i) consider $p(t) = t^4 + t^2 - 1$. This has complex Pisot root $\lambda = i\sqrt{\tau}$, $\tau = \frac{1+\sqrt{5}}{2}$, but $|\lambda| = \sqrt{\tau}$ is not real Pisot. For (ii) consider $q(t) = t^3 - t^2 + 10t - 5$. This has a complex Pisot root λ, but $\omega = |\lambda|^2$ is a root of $r(t) = t^3 - 10t^2 + 5t - 25$, and so is not real Pisot.

Finally, we observe that λ complex Pisot implies that λ is complex Perron. Thus by Theorem 5.16, in each case above, there exists a tiling substitution S with expansion λ. The fact that the corresponding substitution tiling system (X_S, T) is not weakly mixing follows from Remark 7.7, part 3. My thanks to the referee for providing these two examples (see also [**Sol99**]).

The second known mechanism responsible for producing weakly mixing tiling dynamical systems involves quasisymmetry.

PROPOSITION 7.10. [**Rob96b**] *The group of eigenvalues Σ_X of a tiling dynamical system is invariant under the action of the quasicrystallographic point group H_X.*

COROLLARY 7.11. (Radin, [**Rad94**]) *The pinwheel tiling dynamical system is weakly mixing.*

PROOF. Let X be the pinwheel tiling space. Then $\mathbb{T} \subseteq H_X$. Since Σ_X must be discrete and $\mathbb{T}$-invariant, it follows that $\Sigma_X = \{0\}$. □

REMARK 7.12. Since the almost 1:1 extension in Theorem 5.22 is always a metric isomorphism, we can obtain examples (starting e.g. with the binary tilings) of minimal uniquely ergodic finite type tiling spaces that are weakly mixing. On the other hand, the next result shows that none of these examples can be strongly mixing.

THEOREM 7.13. (Solomyak, [**Sol97**]). *No self-affine substitution tiling dynamical system (X, T) can be strongly mixing.*

7.3. Diffraction. X-ray diffraction experiments provide a powerful method for studying the microscopic structure of solids. In particular, quasicrystals were discovered (see [**SBGC84**]) as a result of the observations of unusual diffraction patterns.

Mathematically we model diffraction as follows. A *Delone set* is a uniformly discrete and relatively dense subset of $\mathbb{R}^d$ (see [**Sen95**]). Starting with a tiling x, we let $z = v(x)$ be the Delone set of all its vertex points. We place an "atom" $\delta_{\mathbf{t}}$, consisting of a unit atomic measure, at each vertex point $\mathbf{t} \in z$. The corresponding "solid" consists of the measure

$$\mu_x = \sum_{\mathbf{t} \in v(x)} \delta_{\mathbf{t}}. \tag{7.7}$$

The physical diffraction pattern corresponds mathematically to the locations Σ'_x of the point masses in the Fourier transform $\hat{\mu_x}$. There are many technicalities involving how this is actually computed. For example, one should really compute the transform of the *autocorrelation measure*, which is supported on the difference set $z - z$ of z. We will not discuss these issues here, but rather refer the reader to [**Hof97**] for a nice overview. The main result in this area is the following.

THEOREM 7.14 ([**Dwo93**], [**Hof97**]). *Let (X, T) be a uniquely ergodic tiling space with eigenvalue set Σ. Then for any $x \in X$, $\Sigma'_x \subseteq \Sigma$.*

Because of this result, physicists have been especially interested in tiling dynamical systems with pure discrete spectrum. On the other hand, one would expect a weakly mixing repetitive tiling (like the binary tiling) to have no spots in its diffraction pattern. In spite of this, it still seems reasonable to regard this example as a "quasicrystal": as we will see in the next section, it has entropy zero.

7.4. Entropy.

DEFINITION 7.15. Let (X, T) be a tiling dynamical system. For $n > 0$ the *complexity* $c(n)$ of (X, T) is the number of different equivalence classes of tilings

$x[B_n]$ for all $x \in X$. The *topological entropy* is defined as the exponential growth rate in complexity:

$$h(X) = \lim_{n\to\infty} \frac{1}{n^d} \mathrm{Vol}(B_1)^{-1} \ln(c(n)). \tag{7.8}$$

Nonzero entropy is, in some sense, the ultimate indication of disorder in a dynamical system. A tiling whose orbit closure is a positive-entropy tiling dynamical system should probably be considered too disordered to be regarded as a quasicrystal. Some of the full tiling shifts, discussed above, do have positive entropy. Since they are topologically transitive, there are tilings among them with a positive entropy orbit closure. However, as expected, most of the tiling dynamical systems that arise in the study of quasicrystals do indeed have zero entropy.

THEOREM 7.16. *Suppose X is either* (a) *a substitution tiling space for a primitive invertible tiling substitution, or* (b) *a finite type tiling space which is uniquely ergodic. Then $h(X) = 0$.*

Part (b) is due to Radin [**Rad91**] in the $\mathbb{Z}^d$ case and generalized to the case of tiling dynamical systems by Shieh [**Sh**]. Part (a) follows from the next theorem.

THEOREM 7.17. [**HR02**] *Let (X_S, T) be a substitution tiling space, where $S = LC$ is primitive and invertible. Suppose L has eigenvalues $\lambda_1, \dots, \lambda_d$ where $|\lambda_d| \leq |\lambda_i|$ for all i. Let*

$$c = \frac{\log|\det(L)|}{\log|\lambda_d|} = \frac{\log(|\lambda_1|\cdot \dots \cdot |\lambda_d|)}{\log|\lambda_d|}.$$

Then the complexity satisfies

$$c(n) \leq K \cdot n^c.$$

for some $K > 0$. In the self-similar case, $|\lambda_1| = \dots = |\lambda_d|$, so $c = d$.

The proof follows the dissertation of Clifford Hansen ([**Han00**]), who studied the case of discrete multi-dimensional substitutions. It is based on the following consequence of local finiteness.

LEMMA 7.18. *Let $X_{\mathcal{T}}$ be a finite local complexity tiling space. Given $m \geq 1$ there exists a constant $J = J(m) > 0$ so that for all n sufficiently large*

$$\#\{x \in \mathcal{T}^{(m)} : x \subseteq y \text{ for some } y \in \mathcal{T}^{(n)}\} \leq J \cdot n. \tag{7.9}$$

PROOF OF THEOREM 7.17. For $D_j \in \mathcal{T}$ and $\mathbf{v} = (1, 1, \dots, 1)^t$ we have $\#(C^p D_j) = (A^p \mathbf{v})_j$, where A is the structure matrix for S. Since S is primitive, the Perron-Frobenius Theorem implies

$$\lim_{p\to\infty} \frac{A^p \mathbf{v}}{\omega^p} = (\mathbf{b}\cdot\mathbf{v})\mathbf{a} \stackrel{\text{def}}{=} \mathbf{r},$$

where $\omega > 0$ and $\mathbf{a}, \mathbf{b} > 0$. Since also $\mathbf{v} > 0$, we have $\mathbf{r} > 0$. Thus there exists N so that for all sufficiently large p,

$$\max_{D\in\mathcal{T}} \#(C^p D) \leq N \cdot \omega^p. \tag{7.10}$$

For $p \geq 0$, call a patch $y \in L^p \mathcal{T}^*$ a *p-basic* patch if for some $D \in y$, each $D' \in y$ satisfies $D' \cap D \neq \emptyset$. We denote the p-basic patches, up to equivalence, by $y_1^p, \dots, y_M^p$, where M is independent of p. Then $M' = \max\{\#(y_j^p) : j = 1, \dots, M\}$ is also independent of p.

For a p-basic patch y_j^p, we have $C^p y_j^p \in \mathcal{T}$. Let

$$M_p = \max\{\#(C^p y_j^p) : j = 1, \ldots, M\}. \tag{7.11}$$

Since $\#(C^p y_j^p) \leq M' \cdot \max_{D \in \mathcal{T}} \#(C^p D)$, it follows from (7.10) that

$$M_p \leq M'N \cdot \omega^p. \tag{7.12}$$

Let $\delta = \max\{\mathrm{diam}(D) : D \in \mathcal{T}\}$ and let ϵ be the maximum of all $r > 0$ such that all $D \in \mathcal{T}$ satisfy $B_r \subseteq D - \mathbf{s}$ for some $\mathbf{s} \in \mathbb{R}^d$.

Fix $q \geq (\delta + 1)/\epsilon$. Then for all $n \geq 1$, $\epsilon q n \geq n + \delta$. Define

$$p_n = \frac{\log(qn)}{\log|\lambda_d|}.$$

It follows that

$$\epsilon|\lambda_d|^{p_n} = \epsilon q n \geq n + \delta. \tag{7.13}$$

By the choice of δ, $B_{n+\delta} \supseteq \mathrm{supp}(x[B_n])$ for all $n > 0$ and all $x \in X_S \subseteq X_{\mathcal{T}}$. Since S is invertible, one can define a *super-tiling* $C^{-p}x \in X_{L^p\mathcal{T}}$ for any $p \geq 1$ and $x \in X_S$. It follows from (7.13) that $L^p B_\epsilon \supseteq B_{n+\delta}$ for all $p \geq p_n$. Thus for each $x \in X_S$, the patch $(C^{-p}x)[B_{n+\delta}]$ is a sub-patch of some p-basic patch $T^{\mathbf{t}} y_j^p$ in the super-tiling $C^{-p}x$. Hence,

$$x[B_n] \subseteq x[B_{n+\delta}] = C^p(C^{-p}x[B_{n+\delta}]) \subseteq T^{\mathbf{t}} C^p y_j^p.$$

Now we apply Lemma 7.18, (7.11) and (7.12) to conclude that

$$c(n) \leq J \cdot M_p \leq JM'N \cdot \omega^p \stackrel{\text{def}}{=} K' \cdot \omega^p$$

for all $p \geq p_n$ once n is sufficiently large. In particular, we take n large enough that (7.10) holds for $p = p_n$. It follows that

$$\begin{aligned} c(n) &\leq K'\omega^{p_n} = K'\omega^{\frac{\log(qn)}{\log|\lambda_d|}} \\ &= K' e^{\frac{\log(\omega)\log(qn)}{\log|\lambda_d|}} \\ &= K' q^c \cdot n^c \\ &= K \cdot n^c, \end{aligned}$$

where $K = K'q^c = JM'Nq^c$.

□

We conclude with two open problems.

- Can a uniquely ergodic finite type tiling space X be strongly mixing?
- Is there an example of a uniquely ergodic, finite type tiling space that not an almost 1:1 extension of a substitution tiling space (or, more generally, that is not a tiling space having some other type of hierarchical structure)?

8. Quasiperiodic tilings and models

In this section we describe the *projection method*, which is the third general method (after local matching rules and tiling substitutions) for constructing aperiodic tilings. This method also provides an algebraic/geometric model for the tiling spaces it produces. After describing the projection method, we briefly describe some other geometric models.

8.1. Kronecker dynamical systems. Let $\mathbb{G}$ be a compact abelian group, written additively, with normalized Haar measure γ. Suppose there exists an injective continuous homomorphism $\iota : \mathbb{R}^d \to \mathbb{G}$. The continuous γ-preserving $\mathbb{R}^d$ action

$$T^{\mathbf{t}}\mathbf{g} = \iota(\mathbf{t}) + \mathbf{g} \tag{8.1}$$

on $\mathbb{G}$ is called a *Kronecker dynamical system.* Every point in such a dynamical system is almost periodic. Any coset $W = \overline{\iota(\mathbb{R}^d)} + \mathbf{g}$ is T-invariant. It follows that, up to topological conjugacy, (W, T) is a minimal Kronecker system. Any minimal Kronecker system is uniquely ergodic (see [**Wal82**]).

REMARK 8.1. By an appropriate choice of ι one can ensure that (W, T) is properly minimal.

Every Kronecker system has pure discrete spectrum. The eigenfunctions are the *characters* $\chi \in \widehat{\mathbb{G}}$ of $\mathbb{G}$. In particular, for $\chi \in \widehat{\mathbb{G}}$ we have $\chi(\iota(\mathbf{t})) \in \widehat{\mathbb{R}^d} = \mathbb{R}^d$, so there exists $\mathbf{w} \in \mathbb{R}^d$ such that $\chi(\iota(\mathbf{t})) = e^{\pi i \langle \mathbf{t}, \mathbf{w} \rangle}$, and the set Σ of eigenvalues of (W, T) is the set of all such $\mathbf{w}$ (see [**Wal82**] for details).

Two dynamical systems (X, T) and (Y, T), preserving measures μ and ν respectively, are said to be *metrically isomorphic* (see [**Wal82**]) if there exist T-invariant subsets $X_0 \subseteq X$ and $Y_0 \subseteq Y$ with $\mu(X_0) = \nu(Y_0) = 1$, and an invertible measure preserving Borel mapping $Q : X_0 \to Y_0$ so that $QT = TQ$. Metric isomorphism is the primary notion of isomorphism studied in ergodic theory. If (X, T) and (Y, S) are uniquely ergodic then topological conjugacy implies metric isomorphism. Note that any two metrically isomorphic ergodic dynamical systems must have the same eigenvalues.

HALMOS-VON NEUMANN THEOREM. (See [**Wal82**]) *Any dynamical system with pure discrete spectrum is metrically isomorphic to a Kronecker system. Every countable subgroup $\Sigma \subseteq \mathbb{R}^d$ is the eigenvalue group for a Kronecker system with $\mathbb{R}^d$ acting.*

REMARK 8.2. Every Kronecker system has entropy zero (see [**Wal82**]).

8.2. The projection method. A landmark in the theory of aperiodic tilings is de Bruijn's algebraic theory of Penrose tilings [**dB81**]. Originally, this theory described Penrose tilings as being dual (in the sense of graph theory) to so-called "grid" tilings. The generalization of this idea is called the *grid* method, and the tilings it produces are called *quasiperiodic tilings.* There are two alternate equivalent constructions of quasiperiodic tilings, mostly developed by physicists: the *projection* method and the *cut* method (see [**ODK88**]). Here we discuss the projection method because it is conceptually the simplest.

Let $E^{\|}$ be a d-dimensional subspace of $\mathbb{R}^n$ and let $\iota : \mathbb{R}^n \to E^{\|}$ be an isometric isomorphism. Let $E^{\perp}$ be the perpendicular subspace, so that $\mathbb{R}^n = E^{\|} \oplus E^{\perp}$. Denote the projections to these two subspaces by $\pi^{\|}$ and $\pi^{\perp}$. Consider the integer lattice $\mathbb{Z}^n \subseteq \mathbb{R}^n$. Let W_0 be the closure of $\iota(\mathbb{R}^d)$ in $\mathbb{T}^n = \mathbb{R}^n/\mathbb{Z}^n$ and let $W_0 + \mathbf{g}$, $\mathbf{g} \in E^{\perp}$, be an arbitrary coset. The tiling systems we will construct is closely related to the Kronecker system (W_0, T).

Let $K \subseteq E^{\perp}$ be compact with a Lebesgue measure zero boundary and a nonempty interior. Let $S_K = K + E^{\|}$. For $\mathbf{s} \in \mathbb{R}^d$ let

$$z_{\mathbf{s}} = (\iota^{-1}\pi^{\|})(S_K \cap (\mathbb{Z}^n + \mathbf{s})) \subseteq \mathbb{R}^d.$$

Note that $z_{\mathbf{s}} = z_{\mathbf{s}'}$ if $\mathbf{s} - \mathbf{s}' \in \mathbb{Z}^d$. Thus we can index using $\mathbf{s} \in \mathbb{T}^n = \mathbb{R}^n/\mathbb{Z}^n$, and this makes the mapping $\mathbf{s} \mapsto z_{\mathbf{s}}$ 1:1.

The set $z_{\mathbf{s}}$ is a *Delone set.* It is possible to topologize the collection of all Delone sets (using something similar to a tiling metric) in such a way that the set $\{z_{\mathbf{s}} : \mathbf{s} \in \mathbb{T}^n\}$ is homeomorphic to $\mathbb{T}^n$. We also have $z_{T^{\mathbf{t}}\mathbf{s}} = T^{\mathbf{t}} z_{\mathbf{s}}$, so in fact $\mathbf{s} \mapsto z_{\mathbf{s}}$ is a topological conjugacy.

We call $\mathbf{s} \in \mathbb{T}^n$ *regular* if $\partial(S_K) \cap (\mathbb{Z}^n + \mathbf{s}) = \emptyset$. Let $\mathbb{T}^n_0$ denote the set of all regular points in $\mathbb{T}^n$, and note that $\mathbb{T}^n_0$ has full Lebesgue measure. In many cases it will turn out that the points in $z_{\mathbf{s}}$ are the vertices of a tiling x of $\mathbb{R}^d$. Let us now specialize to such a situation.

Let $d = 2$ and consider the full tiling space $X_{\mathcal{R}_n}$, $n \geq 4$, from Example 2.6. Let $K = \pi^{\perp}(Q)$ where $Q = \{\mathbf{q} \in \mathbb{R}^n : 0 \leq q_i \leq 1\}$ is the unit cube. Let s be as in Example 2.6, and let B be matrix having the vectors $\mathbf{v}_k$ as rows. Define $\iota(\mathbf{t}) = B\mathbf{t}$.

PROPOSITION 8.3 (de Bruijn, [**dB81**]). *For each $\mathbf{s} \in \mathbb{T}^n_0$ there exists a tiling $x \in X_{\mathcal{R}_n}$ with $v(x) = z_{\mathbf{s}}$, i.e., x has the Delone set $z_{\mathbf{s}}$ as its vertex points.*

In addition, de Bruijn [**dB81**] showed that the non-regular points in $\mathbb{T}^n$ correspond to more than one tiling (but always a finite number). Moreover, each of these tilings can be obtained as a limit of the regular cases. This result can be made into a statement about tiling dynamical systems as follows:

THEOREM 8.4. [**Rob96b**] *Let $W = \overline{\iota(\mathbb{R}^2)} + \mathbf{g}$ for $\mathbf{g} \in W^{\perp}$. For $\mathbf{s} \in W \cap \mathbb{T}^n_0$ define $x = H(\mathbf{s}) \in X_{\mathcal{R}_n}$ to be the tiling with vertex set $v(x) = z_{\mathbf{s}}$. Then*

$$X = \overline{H(W \cap \mathbb{T}^n_0)}$$

is a tiling space that is minimal and uniquely ergodic. Moreover, H^{-1} extends to a continuous mapping $P : X \to W$ which is an almost 1:1 factor mapping and a metric isomorphism.

In the case where ι is chosen so that (W, T) is properly minimal, each tiling $x \in X$ will be aperiodic (i.e., properly repetitive). The tilings $x \in X$ are called *quasiperiodic tilings* and (X, T) is called a *quasiperiodic tiling dynamical system.*

COROLLARY 8.5. *Every quasiperiodic tiling dynamical system (X, T) is properly minimal, uniquely ergodic and contains uncountably many incongruent tilings. Moreover, it has an almost 1:1 Kronecker system factor (W, T), to which it is metrically isomorphic. It thus has pure discrete spectrum and entropy zero.*

It is possible to find explicitly the set Σ for these quasiperiodic tiling dynamical systems (see [**Rob96a**]). In particular,

$$\Sigma = B^t \mathbb{Z}^n = \{\sum_{j=1}^{n} n_j \mathbf{v}_j : n_j \in \mathbb{N}\}.$$

When $d = 2$ and $n = 2^m$ we have $\overline{\iota(\mathbb{R}^2)} = \mathbb{T}^n$. The case $n = 4$ gives a well known example called the *octagonal* or *Ammann-Beenker tilings.* Like the Penrose tilings, the Amman-Beenker tilings can also be generated by a local matching rule and by a tiling substitution (see [**Sen95**] for details).

In the case that $d = 2$ and $n = p$ is an odd prime, $\overline{\iota(\mathbb{R}^2)} \cong \mathbb{T}^{p-1}$. Here $E^{\perp}$ is the 1-dimensional subspace generated by $(1, 1, \ldots, 1)$, and $\mathbb{T}^n = W \oplus W^{\perp}$ where

FIGURE 14. A patch of Ammann-Beenker tiling.

$W^\perp = E^\perp/\mathbb{Z}^2 \cong \mathbb{T}$. let us define $\varphi : \mathbb{T}^n \to W^\perp$ by $\varphi(\mathbf{s}) = s_1 + s_2 + \cdots + s_n$ mod 1. For $0 \leq t \leq 1$, we define

$$X_{n,t} = \overline{H((W + \varphi^{-1}(t)) \cap \mathbb{T}^n_0)} \subseteq X_{\mathcal{R}_n}.$$

We can now state de Bruijn's remarkable algebraic structure theorem for Penrose tilings.

THEOREM 8.6. (de Bruijn [**dB81**]) *$X_{5,0}$ is the precisely the set $F(X_{\mathcal{P}})$ of unmarked Penrose tilings.*

COROLLARY 8.7. [**Rob96a**] *The Penrose tiling dynamical system $(X_{\mathcal{P}}, T)$ has pure discrete spectrum, with an almost 1:1 Kronecker factor T on the 4-torus $\mathbb{T}^4$. The eigenvalue group Σ_p is the subgroup of $\mathbb{R}^2$ is generated by the "5th roots of unity" $\mathbf{v}_0, \ldots, \mathbf{v}_4$.*

REMARK 8.8. In a slight abuse of notation we write $\Sigma_p = \mathbb{Z}[e^{2\pi i/5}] \subseteq \mathbb{R}^2$.

The tilings on which the factor map $P : X_{0,5} \to \mathbb{T}^5$ fails to be 1:1 are a well known special class Penrose tilings. In particular, P is 2:1 on the Penrose tilings that contain *infinite worms*, and P is 10:1 on the various *cartwheel tilings*. In particular, $P(x) = 0$ for the cartwheel tilings x centered at the origin. See [**GS87**] for the geometric description of these special Penrose tilings. Similar results hold for all other classes of quasiperiodic tilings.

The cases $X_{5,t}$ for $t \neq 0$, are called *generalized Penrose tilings*. All the generalized Penrose tiling dynamical systems have pure discrete spectrum and they all have the eigenvalue group $\Sigma = \mathbb{Z}[e^{2\pi i/5}]$, which they share with the true Penrose tilings. Thus, by the Halmos von Neumann Theorem, all the corresponding tiling dynamical systems are metrically isomorphic. However, one can also show (see [**Rob96a**], [**Le97**]) that in some cases they are not topologically conjugate. This illustrates the interesting fact that ergodic theory and topological dynamics sometimes provide different invariants when applied to tiling theory.

EXERCISE 10. Show that the point groups satisfy $H_{X_{5,t}} = D_{10}$ only if $t = 0$ or $t = 1/2$, and otherwise $H_{X_{5,t}} = D_5$.

REMARK 8.9. A *Sturmian shift dynamical system* is a symbolic dynamical system obtained by coding an irrational rotation on the circle using a partition into two intervals (see [**Ber00**]). Consider the partition η of a quasiperiodic tiling space X according to which prototile contains the origin. The factor mapping $P : X \to W$ pushes η forward into a partition on W. We can recover the tilings by seeing how this partition tiles the orbits of $\mathbb{R}^2$ in W. In this way, quasiperiodic tiling dynamical systems generalize the idea of Sturmian systems.

8.3. Quasiperiodicity and the finite type property. An interesting general question concerns the relation between quasiperiodic tilings and the finite type property. Such questions were studied primarily by researchers interested in quasicrystals. See [**Le97**] for a good survey with complete references. Here, we discuss mainly the case of generalized Penrose tilings.

THEOREM 8.10. [**Le95**] *Let* $\tau = \frac{1+\sqrt{5}}{2}$.

(1) *The tiling space* $X_{5,t}$ *is a finite type tiling space if and only if* $t \in \mathbb{Z}[\tau]$.
(2) *There exists a finite type tiling space* $X_{\mathcal{T}_\#}$ *by a marked version* $\mathcal{T}_\#$ *of the tiles* $\mathcal{R}_5$ *so that the forgetful mapping* $F : X_{\mathcal{T}_\#} \to X_{5,t}$ *is an almost 1:1 factor if and only if* $t \in \mathbb{Q}(\tau)$. *If* $t \in \mathbb{Q}(\tau)\backslash\mathbb{Z}[\tau]$ *then* F *is not a topological conjugacy; such examples are strictly sofic.*

REMARK 8.11. One should compare this with Theorem 5.22.

REMARK 8.12. In general, a necessary condition for a quasiperiodic tiling system to be finite type or sofic is that the "slope" of $E^{\|}$ must be *algebraic* (in a certain precise sense). Thus sofic examples are very special (see [**Le97**]).

REMARK 8.13. It seems to be still unknown exactly which quasiperiodic tiling systems are substitution tiling systems, and *vice versa*.

8.4. Algebraic and geometric models. The subspace $E^{\perp}$ in the definition of quasiperiodic tiling dynamical systems can be replaced by an arbitrary locally compact abelian group. In this case one obtains more general Delone sets called *model sets*. Such sets can be studied without the benefit of tilings or even dynamical systems theory. However, one can show that quite generally, the corresponding tiling dynamical systems have almost 1:1 Kronecker factors, defined on more general groups than tori. The chair tiling dynamical system can be obtained this way. (See [**BMS98**] for a discussion of this point of view.)

Alternatively, one can ask, given a tiling dynamical system (X, T), whether there is a "classical" dynamical system that is its almost 1:1 factor. We refer to such a factor as an algebraic/geometric *model* for the tiling system. For example the Penrose tilings are modeled by $\mathbb{T}^4$. Here is another example.

Let $\mathbb{G}$ be the locally compact group whose dual $\hat{\mathbb{G}}$ is the group $\mathbb{Z}[\frac{1}{2}] \oplus \mathbb{Z}[\frac{1}{2}]$ of dyadic rational vectors in $\mathbb{R}^2$. By Halmos-von Neumann theory, there is a unique minimal Kronecker $\mathbb{R}^2$ action T on $\mathbb{G}$ with $\Sigma = \mathbb{Z}[\frac{1}{2}] \oplus \mathbb{Z}[\frac{1}{2}]$ (this Kronecker system is called the $\mathbb{R}^2$-*adding machine* in [**Rob99**]).

THEOREM 8.14. [**Rob99**] *The chair tiling dynamical system* (X_{S_c}, T) *has* $(\mathbb{G}, T)$ *as an almost 1:1 factor. Thus it has pure discrete spectrum with* $\Sigma = \hat{\mathbb{G}}$. *The table* *tiling dynamical system* (X_{S_t}, T) *has* $(\mathbb{G}, T)$ *as an almost 4:1 factor. The table system has precisely the same eigenvalues* Σ *as the chair, and thus is not weakly mixing. However, the table tiling system also does not have pure discrete spectrum; it has* mixed spectrum.

REMARK 8.15. The spectral properties of the table and chair were first computed in [**Sol97**], where it was observed that the table system is similar to the well known discrete *Morse sequence* substitution dynamical system (see [**Que87**]).

In the chair (or the table), the points where the factor mapping P fails either to be 1:1 (or 4:1) can be completely classified (see [**Rob99**]). In particular, chair tilings can have worms and cartwheels very similar to the ones that occur in Penrose tilings. The proofs in [**Rob99**] are completely general and show that similar results hold for all polyomino substitutions. In particular, there is an easy criterion for pure discrete spectrum. This turns out to be equivalent to the idea of "coincidence" in the theory of substitutions, and to the idea of "synchronizing" or "magic" words in the theory of discrete 1-dimensional finite type shifts (see [**Rob99**]).

REMARK 8.16. Solomyak [**Sol97**] defines a more general notion of coincidence, and uses it to prove that some examples (e.g., chair tiling systems) have pure discrete spectrum.

Models based on Kronecker systems won't work for weakly mixing examples because weak mixing implies the absence of nontrivial eigenvalues. Thus we ask, what could a model for a weakly mixing tiling dynamical system possibly look like?

8.5. The dynamics of the substitution map and Markov partitions. Before describing an example of a model for a weakly mixing tiling dynamical system, we consider the dynamical properties of the action of an invertible substitution mapping S acting on a tiling space X_S.

THEOREM 8.17. [**Rob96a**] *The Penrose substitution* S *on the set* $X_{\mathcal{P}}$ *of Penrose tilings has the hyperbolic toral automorphism*

$$A = \begin{pmatrix} 0 & 1 & -1 & 0 \\ 1 & 0 & 0 & -1 \\ 0 & 1 & -1 & -1 \\ 1 & 0 & -1 & -1 \end{pmatrix}$$

acting on $\mathbb{T}^4$ *as is an almost 1:1 factor.*

Similarly, one can show that the chair tiling dynamical system has a "hyperbolic" automorphism of $\mathbb{G}$ as an almost 1:1 factor. These examples illustrate that tiling substitutions S tend to be *hyperbolic*.

In [**AP98**] it is shown that the action S of a tiling substitution on X_S (i.e., the dynamical system (X_S, S)) is always a kind of generalized hyperbolic system called a *Smale space*. If the substitution S is based on a perfect decomposition, then then the tilings induce a partition on X_S (according to which prototile occurs at the origin). The perfect decomposition property implies this partition is a *Markov partition* (see [**Bow78**]). Conversely, if a Smale space has a Markov partition whose partition elements are connected, then the partition induces tilings on the stable manifolds, and these tilings are self-affine.

Connectedness is needed because our definition of tilings requires connected tiles. Unfortunately, it is unknown whether every hyperbolic toral automorphism has a Markov partition with connected partition elements. However, some partial results appear in [**FI98**].

REMARK 8.18. It is known that for all Markov partitions for hyperbolic toral automorphisms of $\mathbb{T}^3$ (see [**Bow78**]) and for typical hyperbolic toral automorphisms of $\mathbb{T}^n$, $n > 3$, (see [**Caw91**]) the boundaries of partition elements must be fractal. This implies that self-affine quasiperiodic tilings satisfying a perfect decomposition will almost always have tiles with fractal boundaries.

8.6. A geometric model for a weakly mixing system. The example discussed in this section is based on a kind of hyperbolic dynamical system J called a *pseudo-Anosov diffeomorphism* (see [**FS79**]). In this example, J is defined on a surface M of genus 2. Pseudo-Anosov diffeomorphisms always have Markov partitions, and we obtain self-similar tilings, as in the previous section, by intersecting the Markov elements partition with the unstable manifolds for J. Since for almost every point $m \in M$, the stable manifold through m is homeomorphic to $\mathbb{R}$, this example consists of 1-dimensional tilings. See [**Fit98**] or [**FHR00**] for details.

Consider the discrete substitution σ given by

$$\begin{array}{ccl} 1 & \to & 1424 \\ 2 & \to & 142424 \\ 3 & \to & 14334 \\ 4 & \to & 1434. \end{array} \tag{8.2}$$

The structure matrix for this substitution is

$$A = \begin{pmatrix} 1 & 1 & 1 & 1 \\ 1 & 2 & 0 & 0 \\ 0 & 0 & 2 & 1 \\ 2 & 3 & 2 & 2 \end{pmatrix},$$

which has the non-Pisot Perron-Frobenius eigenvalue $\lambda = \frac{1}{4}(7+\sqrt{5}+\sqrt{2}\sqrt{19+7\sqrt{5}})$. Following example 5.14, we construct the corresponding 1-dimensional tiling substitution S with expansion λ. It follows from Theorem 7.6 that (X_S, T) is weakly mixing.

THEOREM 8.19. [**FHR00**] *Let X_S be the tiling space corresponding to the tiling substitution S in described above. Then (X_S, S) has an almost 1:1 factor that is a pseudo-Anosov diffeomorphism (M, J) on a surface M of genus 2.*

The corresponding tiling dynamical system (X_S, T) is metrically isomorphic a unit speed flow along the unstable manifolds of (M, J). Since not all the stable manifolds are homeomorphic to $\mathbb{R}$, this flow is not defined everywhere on M. However, after removing a "singular set" of measure zero, one can show that this flow is metrically isomorphic to a suspension of a self-inducing interval exchange transformation of four intervals (see [**FHR00**]).

REMARK 8.20. One can obtain a 2-dimensional weakly mixing example by taking the Cartesian square. The geometric model for such an example is a 4-manifold. By Theorem 5.22, this 2-dimensional tiling dynamical system has the property that it is an almost 1:1 factor of a weakly mixing finite type tiling system.

References

[AP98] J. Anderson and I. Putnam, *Topological invariants for substitution tilings and their associated C^* algebras*, Ergodic Thy. and Dynam. Sys. **18** (1998), no. 3, 509–537.

[Brg66] R. Berger, *Undecidability of the domino problem*, Memoirs Amer. Math. Soc. **66** (1966), 72pp.

[Ber00] Valérie Berthé, *Sequences of low complexity: automatic and Sturmian sequences*, Topics in symbolic dynamics and applications (Temuco, 1997), Cambridge Univ. Press, Cambridge, 2000, pp. 1–34.

[BMS98] Michael Baake, Robert V. Moody, and Martin Schlottmann, *Limit-(quasi)periodic point sets as quasicrystals with p-adic internal spaces*, J. Phys. A **31** (1998), no. 27, 5755–5765.

[Bow78] Rufus Bowen, *Markov partitions are not smooth*, Proc. Amer. Math. Soc. **71** (1978), no. 1, 130–132.

[Caw91] Elise Cawley, *Smooth Markov partitions and toral automorphisms*, Ergodic Theory Dynam. Systems **11** (1991), no. 4, 633–651.

[Cul96] Karel Culik, II, *An aperiodic set of* 13 *Wang tiles*, Discrete Math. **160** (1996), no. 1-3, 245–251.

[dB81] N. G. de Bruijn, *Algebraic theory of Penrose's nonperiodic tilings of the plane. I, II*, Nederl. Akad. Wetensch. Indag. Math. **43** (1981), no. 1, 39–52, 53–66.

[Dwo93] Steven Dworkin, *Spectral theory and X-ray diffraction*, J. Math. Phys. **34** (1993), no. 7, 2965–2967.

[FS79] A. Fathi and M. Shub, *Some dynamics of pseudo-anosov diffeomorphisms*, Traveaux de Thurston de les surfaces, Astersique, vol. 66-67, 1979, pp. 181–208.

[Fit98] Thomas Fitzkee, *Weakly mixing tiling flows arising from interval exchange transformations*, Ph.D. thesis, The George Washington University, 1998.

[FHR00] Thomas Fitzkee, Kevin Hockett, and E. Arthur Robinson, Jr., *A weakly mixing tiling dynamical system with a smooth model*, Theoretical Computer Science, **303** (2003), 447-462.

[FI98] Maki Furukado and Shunji Ito, *The quasi-periodic tiling of the plane and Markov subshifts*, Japan. J. Math. (N.S.) **24** (1998), no. 1, 1–42.

[GBN89] F. Girault-Beauquier and M. Nivat, *Tiling the plane with one tile*, Topology and Category Theory in Computer Science, Oxford University Press, New York, 1989, pp. 291–333.

[GL92] C. Godréche and F. Lancon, *A simple example of a non-pisot tiling with five-fold symmetry*, Journal de Physique **2** (1992), 207–220.

[G-S98] Chaim Goodman-Strauss, *Matching rules and substitution tilings*, Ann. of Math. (2) **147** (1998), no. 1, 181–223.

[Got44] W. H. Gottschalk, *Orbit-closure decompositions and almost periodic properties*, Bull. Amer. Math. Soc. **50** (1944), 915–919.

[GS87] Branko Grünbaum and G. C. Shephard, *Tilings and patterns*, W. H. Freeman and Company, New York, 1987.

[Han00] Clifford W. Hansen, *Dynamics of multi-dimensional substitutions*, Ph.D. thesis, The George Washington University, 2000.

[HR02] Clifford W. Hansen and E. Arthur Robinson Jr., *On the complexity of self-affine tilings*, preprint, 2003.

[Hed69] G. A. Hedlund, *Endormorphisms and automorphisms of the shift dynamical system*, Math. Systems Theory **3** (1969), 320–375.

[Hof97] A. Hof, *Diffraction by aperiodic structures*, The mathematics of long-range aperiodic order (Waterloo, ON, 1995), Kluwer Acad. Publ., Dordrecht, 1997, pp. 239–268.

[Hos86] B. Host, *Valeurs propres des systèmes dynamiques définis par des substitutions de longueur variable*, Ergodic Theory Dynam. Systems **6** (1986), no. 4, 529–540.

[Kar96] Jarkko Kari, *A small aperiodic set of Wang tiles*, Discrete Math. **160** (1996), no. 1-3, 259–264.

[Ken90] Richard Kenyon, *Self-similar tilings*, Ph.D. thesis, Princeton University, 1990.

[Ken92] Richard Kenyon, *Rigidity of planar tilings*, Invent. Math. **107** (1992), no. 3, 637–651.

[Ken93] Richard Kenyon, *Erratum: "Rigidity of planar tilings" [Invent. Math.* **107** *(1992), no. 3, 637–651]*, Invent. Math. **112** (1993), no. 1, 223.

[Ken96] Richard Kenyon, *The construction of self-similar tilings*, Geom. Funct. Anal. **6** (1996), no. 3, 471–488.

[Le95] Thang T. Q. Le, *Local rules for pentagonal quasicrystals*, Discrete Comput. Geom. **14** (1995), 13–70.

[Le97] Thang T. Q. Le, *Local rules for quasiperiodic tilings*, The mathematics of long-range aperiodic order (Waterloo, ON, 1995), Kluwer Acad. Publ., Dordrecht, 1997, pp. 331–366.

[Lin84] Douglas Lind, *The entropies of topological Markov shifts and a related class of algebraic integers*, Ergodic Theory Dynam. Systems **4** (1984), no. 2, 283–300.

[LM95] Douglas Lind and Brian Marcus, *An introduction to symbolic dynamics and coding*, Cambridge University Press, Cambridge, 1995.

[Mau89] Christian Mauduit, *Caractérisation des ensembles normaux substitutifs*, Invent. Math. **95** (1989), no. 1, 133–147.

[Mun75] James R. Munkres, *Topology: a first course*, Prentice-Hall Inc., Englewood Cliffs, N.J., 1975.

[ODK88] C. Oguey, M. Duneau and A. Katz, *A geometric approach to quasiperiodic tilings*, Commun. Math. Phys. **118** (1988), 99–118.

[Pen74] Roger Penrose, *Pentaplexy*, Bulletin of the Institute of Mathematics and its Applications **10** (1974), 266–271.

[Pet99] Karl Petersen, *Factor maps between tiling dynamical systems*, Forum Math. **11** (1999), no. 4, 503–512.

[PS01] N. Priebe and B. Solomyak, *Characterization of planar pseudo-self-similar tilings*, Discrete Comput. Geom. **26** (2001), no. 3, 289–306.

[Que87] Martine Queffélec, *Substitution dynamical systems—spectral analysis*, Lecture Notes in Mathematics, vol. 1294, Springer-Verlag, Berlin, 1987.

[Rad91] Charles Radin, *Disordered ground states of classical lattice models*, Rev. Math. Phys. **3** (1991), no. 2, 125–135.

[Rad94] Charles Radin, *The pinwheel tilings of the plane*, Ann. of Math. (2) **139** (1994), no. 3, 661–702.

[Rad99] Charles Radin, *Miles of tiles*, American Mathematical Society, Providence, RI, 1999.

[RW92] Charles Radin and Mayhew Wolff, *Space tilings and local isomorphism*, Geom. Dedicata **42** (1992), no. 3, 355–360.

[RS01] Charles Radin and Lorenzo Sadun, *Isomorphism of hierarchical structures*, Ergodic Theory Dynam. Systems **21** (2001), no. 4, 1239–1248.

[rRob71] Raphael M. Robinson, *Undecidability and nonperiodicity of tilings of the plane*, Inventiones Math. **12** (1971), 177–209.

[Rob96a] E. Arthur Robinson, Jr., *The dynamical properties of Penrose tilings*, Trans. Amer. Math. Soc. **348** (1996), no. 11, 4447–4464.

[Rob96b] E. Arthur Robinson, Jr., *The dynamical theory of tilings and quasicrystallography*, Ergodic theory of $\mathbf{Z}^d$ actions (Warwick, 1993–1994), London Math. Soc. Lecture Note Ser., vol. 228, Cambridge Univ. Press, Cambridge, 1996, pp. 451–473.

[Rob99] E. Arthur Robinson, Jr., *On the table and the chair*, Indag. Math. (N.S.) **10** (1999), no. 4, 581–599.

[Rud88] Daniel J. Rudolph, *Rectangular tilings of $\mathbb{R}^n$ and free $\mathbb{R}^n$-actions*, Dynamical systems (College Park, MD, 1986–87), Springer, Berlin, 1988, pp. 653–688.

[Rud89] Daniel J. Rudolph, *Markov tilings of $\mathbb{R}^n$ and representations of $\mathbb{R}^n$ actions*, Measure and measurable dynamics (Rochester, NY, 1987), Amer. Math. Soc., Providence, RI, 1989, pp. 271–290.

[Rue69] David Ruelle, *Statistical mechanics: Rigorous results*, W. A. Benjamin, Inc., New York-Amsterdam, 1969.

[Sal63] Raphaël Salem, *Algebraic numbers and Fourier analysis*, D. C. Heath and Co., Boston, Mass., 1963.

[SBGC84] D. Schechtman, I. Blech, D. Gratias, and J. W. Cahn, *Metallic phase with long range orientational order and no translational symmetry*, Phys. Rev. Letters **53** (1984), 1951–1953.

[Sen95] Marjorie Senechal, *Quasicrystals and Geometry*, Cambridge University Press, Cambridge, 1995.

[Sh] Jiunn-I Shieh, *The entropy of uniquely ergodic tiling systems*, preprint, University of Texas.

[Sol97] Boris Solomyak, *Dynamics of self-similar tilings*, Ergodic Theory Dynam. Systems **17** (1997), no. 3, 695–738.

[Sol98] B. Solomyak, *Nonperiodicity implies unique composition for self-similar translationally finite tilings*, Discrete Comput. Geom. **20** (1998), no. 2, 265–279.

[Sol99] Boris Solomyak, *Corrections to: "Dynamics of self-similar tilings" [Ergodic Theory Dynam. Systems* **17** *(1997), no. 3, 695–738]*, Ergodic Theory Dynam. Systems **19** (1999), no. 6, 1685.

[Thu89] W. Thurston, *Groups, tilings and finite state automata*, AMS Colloquium Lecture Notes, Boulder, Colorado, 1989.

[Wal82] Peter Walters, *An introduction to ergodic theory*, Springer-Verlag, New York, 1982.

[Wan61] H. Wang, *Proving theorems by pattern recognition. ii.*, Bell System Technical Journal **40** (1961), 1–42.

(E.A. ROBINSON) DEPARTMENT OF MATHEMATICS, GEORGE WASHINGTON UNIVERSITY, WASHINGTON, DC 20052., USA

E-mail address: `robinson@gwu.edu`

Proceedings of Symposia in Applied Mathematics
Volume **60**, 2004

Strong Shift Equivalence Theory

J.B.Wagoner

ABSTRACT. Strong shift equivalence theory arose from the long standing classification problem in symbolic dynamics. We will discuss how it is closely related to areas of mathematics outside dynamics such as algebraic K-theory, cyclic homology, and topological quantum field theory.

1. Introduction

The main goal of this paper is to explain the diagram below which summarizes the history of strong shift equivalence theory, starting at the bottom with R.F Williams' foundational work on the classification problem.

In Section 2 we state the classification problem for subshifts of finite type and discuss the concept of strong shift equivalence in elementary terms using matrices and directed graphs. Section 3 introduces the strong shift equivalence spaces $SSE(\Lambda)$ and $SSE(\Lambda^+)$ for a ring Λ, and discusses their relationship to automorphisms of shift spaces and dimension groups. Section 4 and Section 5 discuss subshifts and their automorphisms from the viewpoint of nonnegative polynomial matrices. Section 6 explains the Δ-strategy for finding counterexamples to the Shift Equivalence Problem. Section 7 brings together the SSE spaces and the polynomial method to construct a K_2 invariant for strong shift equivalence. Section 8 gives a concrete counterexample to the Shift Equivalence Problem. Section 9 is about the conjectural connection between the large group of inert automorphisms of a shift, the algebraic K-theory group K_3, and the cyclic homology Chern character. Finally, Section 10 discusses a relationship between SSE and topological quantum field theory. Some helpful general references are [K,LM,R].

1991 *Mathematics Subject Classification.* Primary 19C99, 19D55, 58F99, 81R99.
Key words and phrases. Strong shift equivalence, the shift equivalence problem, nonnegativity.
The author was supported in part by NSF Grant DMS 9971501.

SSE Theory

$\downarrow$

$\pi_2(SSE(Z), A) = 0$

$\downarrow$

$\pi_2(SSE(Z), SSE_{2m}(Z^+)) \overset{?}{\longrightarrow} K_3(Z[t]/(t^{m+1}), (t)) \longrightarrow Q$

$\| \quad Inert(\sigma_A)/Simp(\sigma_A)$ — Cyclic homology Chern character for $m = 1$

$\downarrow$

$\pi_1(SSE_{2m}(Z^+), A) = Aut(\sigma_A)/Simp(\sigma_A)$

δ_A — Krieger's dimension group representation

$\downarrow$

$\pi_1(SSE(Z), A) = Aut(s_A)$

$\downarrow$

$\pi_1(SSE(Z), SSE_{2m}(Z^+)) \xrightarrow{\Phi_{2m}} K_2(Z[t]/(t^{m+1}), (t)) = Z/2Z$

1970, computation by van der Kallen when $m = 1$

$\downarrow$

$\pi_0(SSE_{2m}(Z^+))$

$\downarrow$ — Injective ? 1974, Williams
No ! 1997, Kim-Roush using sgc_2
1998, Author using K_2

$\pi_0(SSE(Z))$

2. The Classification Problem

We first discuss the still open classification problem and its relation to the concept of strong shift equivalence introduced by R.F. Williams' in [Wi1].

The *full m-shift* is the set X_m of bi-infinite sequences $x = \{x_k\}$ where each x_k comes from a set S of symbols of cardinality m. Typically, $S = \{0, 1, \ldots, m-1\}$. It is equipped with the product topology making it a Cantor set. The shift homeomorphism $\sigma_m : X_m \to X_m$ is defined by $\sigma_m(x)_k = x_{k+1}$. An $m \times m$ zero-one matrix A determines a *subshift of finite type* $\{X_A, \sigma_A\}$ of $\{X_m, \sigma_m\}$ by letting X_A be the subspace of sequences satisfying $A(x_k, x_{k+1}) = 1$ for all $-\infty < k < \infty$. A is the *transition matrix*, and we let $\sigma_A = \sigma_m | X_A$.

More generally, the standard *edge shift* construction [LM] allows one to obtain a subshift of finite type (X_A, σ_A) from any $m \times m$ nonnegative integral matrix $A = \{A_{ij}\}$. The matrix A can be viewed as a directed graph which has m vertices and which has A_{ij} edges going from the vertex i to the vertex j. Order the set $S^\#$ of edges of the graph A and define the zero-one matrix $A^\# : S^\# \times S^\# \to \{0,1\}$ by letting $A^\#(\alpha, \beta) = 1$ iff the end vertex of the edge α is the start vertex of the edge β. Then one defines

$$(X_A, \sigma_A) = \{X_{A^\#}, \sigma_{A^\#}\} \tag{2.1}$$

This is a subshift of the full Bernoulli $m^\#$-shift where $m^\#$ is the number of edges of A. For example, $(X_2, \sigma_2) = (X_A, \sigma_A)$ where $A = \{2\}$, because $A^\#$ is the 2×2 matrix of all 1's. If A is a zero-one matrix, the shifts $\{X_A, \sigma_A\}$ and $\{X_{A^\#}, \sigma_{A^\#}\}$ are cannonically equivalent as explained in [W4].

A reversible discrete dynamical system is a typically represented as a homeomorphism $f : X \to X$. Suppose we have written X as a disjoint union of sets $X = U_1 \cup U_2 \cup \ldots \cup U_n$. Define $\pi : X \to X_n$ as follows: $\pi(x)_k = i$ iff $f^k(x)$ is in U_i. Then $\pi f = \sigma_n \pi$. If $\{U_1, \ldots, U_n\}$ happens to be a Markov partition, then π is a homeomorphism and is therefore a *topological conjugacy* between (X, f) and the subshift of finite type $\{X_A, \sigma_A\}$ where $A(i,j) = 1$ iff $U_i \cap f^{-1}U_j$ is not empty. The standard example of a Markov partition for (X_A, σ_A) coming from an $r \times r$ zero-one matrix A is $\{U_1, U_2, \ldots, U_r\}$ where U_i consists of those sequences $x = \{x_k\}$ in X_A where $x_0 = i$. However, if one Markov partition exists for (X, f), then there are infinitely many. So we have the fundamental and still open

Classification Problem *Given nonnegative integral matrices A and B, under what conditions are (X_A, σ_A) and (X_B, σ_B) topologically conjugate?*

Dynamical systems which are topologically conjugate have the same underlying behaviour. In this regard subshifts of finite type are systems where it is relatively easy to explicitly compute dynamically intrinsic quantities. See [LM]. One of the most important and basic invariant quantities is the *topological entropy* $h(f)$ defined by Adler-Konheim-McAndrew in [AKM]. This is a topological version of entropy in ergodic theory [Kol] and channel capacity in information theory [Sh]. When $X = X_A$ and $f = \sigma_A$, the topological entropy $h(\sigma_A)$ can be computed by the formula

$$h(\sigma_A) = lim_{n \to \infty} \frac{1}{n} \log B_n = log\lambda$$

where B_n is the number of blocks of length n appearing in sequences of X_A and where $\lambda = \lambda_A$ is the Perron-Frobenius eigenvalue, the largest real root of $det(tI - A)$. See [LM, Section 4]. Another important invariant is the Artin-Mazur *zeta function* which elegantly combines information about the number of periodic points. For a subshift of finite type it is given by the formula

$$\zeta_A(t) = exp(\sum_{n=1}^{\infty} \frac{p_n(\sigma_A)}{n} t^n)$$

where $p_n(\sigma_A)$ is the number of periodic points x in X_A such that $\sigma_A^n(x) = x$. In fact, $p_n(\sigma_A) = Tr(A^n)$, and there is the Bowen-Lanford formula [LM, 6.4.6] which says that

$$\zeta_A(t) = \frac{1}{det(I - tA)}.$$

While the entropy and zeta functions are useful invariants of subshifts, they are certainly not enough for classification. R.F. Williams formulated a general algebraic approach this problem, which we now review. An *elementary strong shift equivalence* $(R, S) : M \to N$ over the nonnegative integers Z^+ consists of two nonnegative integral matrices of finite size R and S satisfying the strong shift equivalence equations

$$\text{(SSE)} \qquad M = RS \quad , \quad SR = N$$

Williams defined the matrices A and B to be *strong shift equivalent over* Z^+ iff there is a chain of elementary strong shift equivalences over Z^+ between them. We shall see below that R and S play different roles, so the direction of the arrow in the notation $(R, S) : M \to N$ is important. In [Wi1] Williams proved

THEOREM 2.2. *(X_A, σ_A) and (X_B, σ_B) are topologically conjugate iff A and B are strong shift equivalent over Z^+.*

Also see [Fr,LM,W1] for expositions of this result. Here is an example. Let

$$A = \begin{pmatrix} 1 & 1 \\ 1 & 1 \end{pmatrix} \qquad B = \begin{pmatrix} 1 & 1 & 0 \\ 0 & 0 & 1 \\ 1 & 1 & 1 \end{pmatrix}$$

$$R = \begin{pmatrix} 1 & 1 & 0 \\ 0 & 0 & 1 \end{pmatrix} \qquad S = \begin{pmatrix} 1 & 0 \\ 0 & 1 \\ 1 & 1 \end{pmatrix}$$

Observe that $\{X_A, \sigma_A\} = (X_2, \sigma_2)$ and that $\{X_B, \sigma_B\}$ is a subshift of (X_3, σ_3). We have $A = RS$ and $SR = B$, and the strong shift equivalence $(R, S) : A \to B$ produces a topological conjugacy

$$\text{(2.3)} \qquad c(R, S) : \{X_A, \sigma_A\} \to \{X_B, \sigma_B\}$$

as follows. Let $x = \{x_k\}$ and $y = \{y_k\}$ where $y = c(R, S)(x)$. Then

$$1 = A(x_k, x_{k+1}) = \sum_i R(x_k, i)S(i, x_{k+1}).$$

Since A, R, and S are zero-one matrices, there is exactly one i for which $R(x_k, i) = S(i, x_{k+1}) = 1$. By definition, $y_k = i$. We have $c(R, S)c(S, R) = c(A, 1) = \sigma_A$ and $c(S, R)c(R, S) = c(B, 1) = \sigma_B$. Here composition is read from left to right. This process generalizes to matrices over Z^+ using the standard edge path construction. See [LM,7.2] and [W4]. See also Section 5 below.

Strong shift equivalence over Z^+ is difficult to determine. A more algebraically tractable concept is shift equivalence. The matrices A and B are said to be *shift equivalent* over Z^+ iff there are nonnegative matrices R and S and a positive integer k, called the *lag*, satisfying the shift equivalence equations

(SE) $AR = RB \quad , \quad BS = SA \quad , \quad A^k = RS \quad , \quad SR = B^k$

Two subshifts of finite type (X_A, σ_A) and (X_B, σ_B) are *eventually conjugate* if there is an integer N such that (X_A, σ_A^k) and (X_B, σ_B^k) are topologically conjugate for $k \geq N$. Williams [Wil] and Kim-Roush [KR1] proved

THEOREM 2.4. *(X_A, σ_A) and (X_B, σ_B) are eventually conjugate iff A and B are shift equivalent over Z^+.*

See[LM, 7.5.15]. Obviously, conjugacy implies eventual conjugacy and strong shift equivalence implies shift equivalence. Williams' influential work [Wil] brought forth the following question.

SHIFT EQUIVALENCE PROBLEM 2.5. *Does SE over Z^+ imply SSE over Z^+ ?*

Shift equivalence over Z^+ is much easier to determine than strong shift equivalence over Z^+. Shift equivalence is decidable [KR1,KR2], and Krieger characterized shift equivalence in terms of the *dimension group* triple (G_A, G_A^+, s_A). See [LM,7.5]. The most concrete definition of G_A uses the eventual range R_A. Suppose A is an $n \times n$ matrix and consider it as a linear transformation from Q^n to itself. We let $R_A = A^m(Q^n)$ where m is sufficiently large that $A : A^m(Q^n) \to A^{m+1}(Q^n)$ is an isomorphism. Then

$$\begin{aligned} &G_A = \{v| \ \ v\epsilon R_A \quad \text{and} \quad vA^p \epsilon Z^n \quad \text{for some integer } p \geq 0 \ \} \\ (2.6) \qquad &G_A^+ = \{v| \ \ v\epsilon G_A \quad \text{and } vA^p \geq 0 \text{ for some integer } p \geq 0 \ \} \\ &s_A \text{ is the isomorphism of } G_A \text{ induced by } A \end{aligned}$$

Example 2.7. $A = \{2\}$. Then $G_A = Z[\frac{1}{2}]$, $G_A = Z[\frac{1}{2}]^+$, and s_A is multiplication by 2.

Example 2.8. Consider the golden mean shift defined by the matrix

$$A = \begin{pmatrix} 1 & 1 \\ 1 & 0 \end{pmatrix}$$

Since $det(A) = -1$, we have $G_A = Z^2$. The characteristic polynomial of A is $t^2 - t - 1$ and the Perron-Frobenius eigenvalue λ satisfies the equation $\lambda^2 = \lambda + 1$. The vector $(\lambda, 1)$ is a right eigenvector for λ. The correspondence sending the pair of integers (a, b) to $a\lambda + b$ induces an isomorphism between G_A and $Z[\lambda]$, considered as a subgroup of the real numbers R equipped with its usual ordering. Via this isomorphism, s_A corresponds to multiplication by λ and G_A^+ becomes $Z[\lambda]^+$.

The name "dimension group" comes from operator algebra theory where it is possible to have dimension functions on the projections in infinite dimensional algebras which may take on nonintegral real number values. In particular, a nonnegative integral matrix A gives rise to a certain C*-algebra via a Bratelli diagram so that the range of the dimension function is G_A. See [E,W6] for expositions of these ideas.

Krieger observed the following useful result. See [LM,7.5.8].

THEOREM 2.9. *Nonnegative integral matrices A and B are shift equivalent over Z^+ iff (G_A, G_A^+, s_A) is isomorphic to (G_B, G_B^+, s_B).*

The most important case is when A and B are *primitive.* This means there is a positive integer m such that both A^m and B^m have all entries positive. The dynamical significance is that the subshifts (X_A, σ_A) and (X_B, σ_B) are topologically mixing [LM, Section 6.3]. It turns out that primitive matrices are shift equivalent over Z^+ iff they are shift equivalent over Z [LM, 7.5]. This implies that primitive A and B are shift equivalent over Z^+ iff the dimension group pairs (G_A, s_A) and (G_B, s_B) are isomorphic. See [LM,7.5.9]. When $det(A) = \pm 1$, there is the very concrete description

$$(2.10) \qquad G_A = Z^n, \qquad s_A = \text{the isomorphism of } G_A \text{ given by } A$$

The above discussion yields the

COROLLARY 2.11. *Suppose $det(A) = \pm 1$ and $det(B) = \pm 1$. Then A and B are shift equivalent over Z^+ iff the matrices A and B are conjugate in $Gl_n(Z)$.*

In certain categories SE does imply SSE. Effros-Williams showed that SE over Z implies SSE over Z. For an exposition of this, see [W3] or see [BH1] where a proof is given that SE implies SSE for Dedekind domains Λ. In general, there is the

ALGEBRAIC SHIFT EQUIVALENCE PROBLEM 2.12. *For what rings Λ does SE over Λ imply SSE over Λ ?*

In any case, when $\Lambda = Z$ we know that SE over Λ^+ is equivalent to SSE over Λ for primitive matrices. So an alternative statement of the Shift Equivalence Problem over Z^+ is therefore

STRONG SHIFT EQUIVALENCE PROBLEM 2.13. *If A and B are primitive non-negative integral matrices, does SSE over Z imply SSE over Z^+ ?*

This reformulation turns out to be a very fruitful new viewpoint, because it recasts the Shift Equivalence Problem into a question of studying SSE over Z versus SSE over Z^+. The answer to (2.13) is sometimes "yes" for 2×2 matrices [B,CK1,Wi2], although here is a well known example where the question is still open. Let

$$(2.14) \quad A_n = \begin{pmatrix} 1 & n(n-1) \\ 1 & 1 \end{pmatrix} \quad B_n = \begin{pmatrix} 1 & n \\ n-1 & 1 \end{pmatrix} \quad R_n = \begin{pmatrix} n-1 & n \\ 1 & 1 \end{pmatrix}$$

Then $det(R_n) = -1$ and $A_n R_n = R_n B_n$. So A_n and B_n are SSE over Z. Observe that $A_2 = B_2$. Kirby Baker [B] showed that A_3 is SSE to B_3 over Z^+. There is a chain of seven elementary SSE's between them, some of which involve 4×4 matrices. See [LM, p.238]. Is A_n SSE to B_n over Z^+ for $n \geq 4$? Kim-Roush have shown in [KR7,KR8] that A_n is SSE to B_n over Q^+ for $n \geq 2$.

A specific 7×7 counterexample to the Strong Shift Equivalence Problem similar to the first such example given in [KR5,KR6] will be discussed in Section 8.

3. Strong shift equivalence spaces

A general program for studying the difference between SSE over Z^+ and SSE over Z involves the *strong shift equivalence spaces* $SSE(Z^+)$ and $SSE(Z)$. These CW complexes arose in the study of automorphism groups of subshifts of finite type. See [BaW, W1-W5, W7, KRW1].

The vertices of $SSE(Z^+)$ are finite, square matrices A with nonnegative integer entries. The edges are elementary strong shift equivalences $(R, S) : A \to B$ over Z^+. It is possible that $A = B$, in which case a loop is created at the vertex A. The 2-cells come from triangles

(3.1)
$$\begin{array}{ccccc} & & B & & \\ & {\scriptstyle (R_1,S_1)} \nearrow & & \searrow {\scriptstyle (R_2,S_2)} & \\ A & & \xrightarrow[(R_3,S_3)]{} & & C \end{array}$$

where the following *Triangle Identities* hold:

(3.2)
$$R_1R_2 = R_3,\ R_2S_3 = S_1,\ S_3R_1 = S_2$$

The precise definition of an n-cell of $SSE(Z^+)$ is given in [W2,W4] and at the end of Section 9 below. The definition of $SSE(Z)$ is entirely similar, as is the definition of $SSE(ZO)$ which is formed by considering only matrices in the category ZO of zero-one matrices. Ditto for $SSE(\Lambda^+)$ and $SSE(\Lambda)$ for more general rings Λ. Strictly speaking, in forming $SSE(ZO)$ or $SSE(\Lambda^+)$ we assume the technical condition that each row and each column of a vertex A has a nonzero entry. The case when Λ is a ring of integral Laurent polynomials arises from Markov chains. Λ^+ consists of those Laurent polynomials with nonnegative coefficients. See [Bo1-Bo3, MT1-MT3, T]. The case when Λ is the group ring $Z[G]$ of a finite group G arises from free actions of G on subshifts of finite type. The motivation for this construction is discussed in [W7], and a guiding analogy is that the space of Markov partitions on a subshift of finite type plays a role similar to the space of C^∞ functions on a smooth manifold.

It follows from the definitions and Williams paper [Wil] that

(3.3)
$$\begin{aligned} \pi_0(SSE(ZO)) &= \textit{strong shift equivalence classes over } ZO \\ &\qquad\qquad\qquad \| \\ \pi_0(SSE(Z^+)) &= \textit{strong shift equivalence classes over } Z^+ \\ \\ \pi_0(SSE(Z)) &= \textit{strong shift equivalence classes over } Z \end{aligned}$$

See [W4,2.1] also.

The group $Aut(\sigma_A)$ of *automorphisms* of the shift (X_A, σ_A) consists of those homeomorphisms $f : X_A \to X_A$ which commute with σ_A. It was first studied by Hedlund and coworkers at IDA in the 1960's in connection with coding theory. See[H]. In the 1980's there was a renewal of interest in $Aut(\sigma_A)$. See [BLR]. Clearly, σ_A itself belongs to $Aut(\sigma_A)$. Any permutation α on n symbols gives an element of $Aut(\sigma_n)$ by simply permuting each symbol; i.e., $\alpha(x)_k = \alpha(x_k)$ for $x = \{x_k\}$.

$Aut(\sigma_n)$ contains the direct sum of any countable set of finite groups as well as the direct sum of a countable number of copies of Z. In general, $Aut(\sigma_A)$ is a huge noncommutative countable group when A is primitive and not equal to the one-by-one matrix $\{1\}$. Other work on $Aut(\sigma_A)$ may be found in [BF, BK1, BK2, F, KRW1, KRW2, KRW3, N, W1-W5]. An important normal subgroup of $Aut(\sigma_A)$ is the group $Simp(\sigma_A)$ of *simple* automorphisms [N], which arise from automorphisms of graphs that keep the vertices fixed. Consider a directed graph constructed from a matrix P over Z^+ where there are P_{kl} arcs going from the vertex k to the vertex l as in the diagram

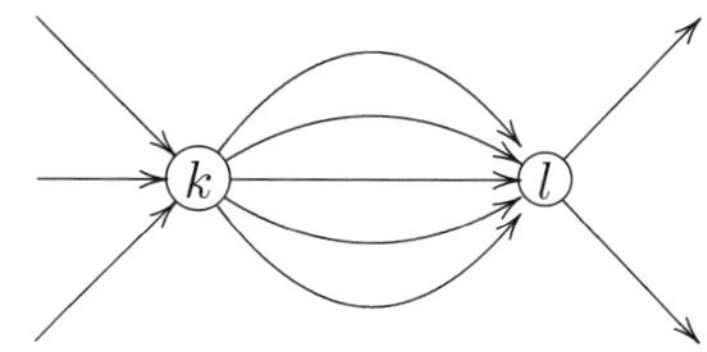

with, for example, $P_{kl} = 5$. Any permutation α of the arcs from k to l induces a simple automorphism α of (X_P, σ_P). By definition, $Simp(\sigma_A)$ is generated by automorphisms of (X_A, σ_A) which are conjugate to those like α for various graphs P.

We let $Aut(s_A)$ denote the group of *automorphisms of the dimension pair* (G_A, s_A). By definition, this consists of those automorphisms α of the group G_A which commute with the isomorphism s_A. We do not require α to preserve the nonnegative elements G_A^+. When $det(A) = \pm 1$, we have the very concrete description

(3.4) $\quad Aut(s_A) = \text{the elements of } Gl_n(Z) \text{ which commute with } A$

If, in addition, the characteristic polynomial $det(tI - A)$ is irreducible, we have

(3.5) $$Z[A]^* \subset Aut(s_A) \subset O^* \subset Q[A]^*$$

where O is the ring of integers in the number field $Q[\lambda] \simeq Q[A]$ generated by the Perron-Frobenius eigenvalue λ of A. The notation R^* means the group of invertible elements in a ring R. See [LM, BLR, BMT]. In particular, using (3.5) it is possible in many cases to explicitly compute generators for $Aut(s_A)$ using the computational algebra program PARI.

THEOREM 3.6. *There are isomorphisms*

(A)
$$\pi_1(SSE(ZO), A) = Aut(\sigma_A)$$
$$\pi_i(SSE(ZO), A) = 0 \qquad \textit{for } i \geq 2$$

(B)
$$\pi_1(SSE(Z^+), A) = Aut(\sigma_A)/Simp(\sigma_A)$$

(C)
$$\pi_1(SSE(Z), A) = Aut(s_A)$$
$$\pi_i(SSE(Z), A) = 0 \qquad \textit{for } i \geq 2$$

This is proved in [BaW, W1-W4].

This was very encouraging from the viewpoint of topology, because it said that $SSE(ZO)$ is a construction of the classifying space of the discrete group $Aut(\sigma_A)$ which is quite different from the universal bar construction method in homological algebra. Such circumstances have historically tended to yield new information, and that turned out to be the case in this subject too.

An important representation of $Aut(\sigma_A)$ is the *dimension group homomorphism*

$$\delta_A : Aut(\sigma_A) \to Aut(s_A) \tag{3.7}$$

which was first defined dynamically by Kreiger as explained in [BLR, LM, W6]. In view of (3.6), it also can be defined using the homomorphism of fundamental groups

$$\delta_A : \pi_1(SSE(ZO), A) \to \pi_1(SSE(Z^+), A) \to \pi_1(SSE(Z), A) \tag{3.8}$$

The image of δ_A actually lies inside the subgroup of automorphisms of (G_A, s_A) which preserve G_A^+, although we will not make use of this. If $(R, S) : A \to B$ over ZO or Z^+, then the induced isomorphism of dimension group triples

$$g(R) : (G_A, G_A^+, s_A) \to (G_B, G_B^+, s_B)$$

just comes from the isomorphism from the eventual range R_A of A to the eventual range R_B of B induced by R. Ditto for $(R, S) : A \to B$ over Z, except that G_A^+ does not necessarily go to G_B^+. Suppose γ in $Aut(\sigma_A)$, $Aut(\sigma_A)/Simp(\sigma_A)$, or $Aut(s_A)$ is represented as a loop

$$\gamma = \prod_{i=1}^{m} \gamma(R_i, S_i)^{\epsilon_i} \tag{3.9}$$

over ZO, Z^+, or Z respectively where $\gamma(R_i, S_i)$ is the path between A_{i-1} and A_i corresponding to the strong shift equivalence (R_i, S_i) and where $\epsilon_i = +1$ if $(R_i, S_i) : A_{i-1} \to A_i$ and $\epsilon_i = -1$ if $(R_i, S_i) : A_i \to A_{i-1}$. Then using (C) in (3.6) we have

$$\delta_A(\gamma) = \prod_{i=1}^{m} g(R_i)^{\epsilon_i} \tag{3.10}$$

A key step in proving (3.6) is

THEOREM 3.11. *Let $\gamma : \{X_A, \sigma_A\} \to \{X_B, \sigma_B\}$ be a topological conjugacy where A and B are in ZO. Then there is a chain of elementary strong shift equivalences $(R_1, S_1), \ldots, (R_m, S_m)$ between A and B in ZO such that*

$$\gamma = \prod_{i=1}^{m} c(R_i, S_i)^{\epsilon_i}$$

where $\epsilon_i = +1$ if $(R_i, S_i) : A_{i-1} \to A_i$ and $\epsilon_i = -1$ if $(R_i, S_i) : A_i \to A_{i-1}$.

There are many ways in which γ can be written as such a product corresponding to different "paths" of elementary strong shift equivalences connecting A and B. It is natural to ask if there are some general relations between the various $c(R, S)$ which would correspond to deformations or homotopies between paths connecting A and B. Put slightly differently, is there a natural notion of deformation or homotopy between paths connecting A and B such that all homotopic paths give the same γ ? And does this notion of homotopy capture all relations between the $c(R, S)$? The affirmative answer is (3.6) and involves the Triangle Relations which came from studying the contractible space of Markov partitions. See [W7].

The main point in the proof of Williams' classification theorem really amounts to showing that the Markov partition space is connected. The key point in proving (3.6) is simple connectivity of the Markov partition space. See [BaW,Fr,W1,Wi1].

4. Nonnegative Polynomial Matrices

Matrices with nonnegative polynomial entries provide a very compact, efficient, and powerful way of representing subshifts of finite type. Perhaps the first appearance of this idea is in Shannon's work [Sh] on information theory. In [KRW2,KRW3] and [KOR] the polynomial matrix technique is indispensible in studying automorphisms of shift spaces on the one hand and characterizing the spectra of primitive, nonnegative integral matrices on the other. See [BW] for basic material and [Bo2] for a survey.

Consider a *nonnegative polynomial matrix* $A = \{A_{ij}\}$ where $A_{ij} = A_{ij}(t)$ lies in $Z^+[t]$, the set of polynomials in t with nonnegative integer coefficients. The indices i and j will range through the positive integers, and we will assume that A has *finite support*, i.e., $A_{ij} \neq 0$ for at most finitely many pairs of indices (i,j). We will generally let I denote the identity matirx of infinite size and make the blanket assumption that all other matrices will have finite support. This section describes how to construct a subshift of finite type $(X(A), \sigma(A))$ under the *no Z-cycles condition (NZC)*

(4.1) The nonnegative integer matrix $A(0) = \{A_{ij}(0)\}$ has no periodic cycles

This construction generalizes the one in [KRW2,KRW3] where it was assumed that $A(0) = 0$, i.e., each polynomial A_{ij} is divisible by t. From the viewpoint of this paper, a compelling reason to construct $(X(A), \sigma(A))$ in the presence of NZC is to give geometric meaning in the context of symbolic dynamics to the Polynomial Strong Shift Equivalence equations in Section 7. These are similar to equations found in algebraic K-theory.

The first step is to construct the directed graph $A^\flat$ as follows: The indices i and j will be called the *primary vertices.* Suppose

$$A_{ij} = a_0 + a_1 t + a_2 t^2 + \ldots + a_n t^n$$

The coefficient a_0 will be called a *constant term* and $a_k t^k$ for $k > 0$ will be called a *power term.* Corresponding to the constant term a_0 in A_{ij} draw a_0 arcs from i to j. These will be called *constant term routes.* Corresponding to the power term $a_k t^k$ in A_{ij} draw a_k simple paths of length k from i to j, each having k edges and $k-1$ *secondary vertices.* These paths from i to j will be called *power term routes.* In particular, each route intersects the set of primary vertices only in its starting vertex and its ending vertex. For example, the matrix

$$A = \begin{pmatrix} t + t^2 & 1 + 2t^3 \\ t & 0 \end{pmatrix}$$

gives rise to the graph $A^\flat$

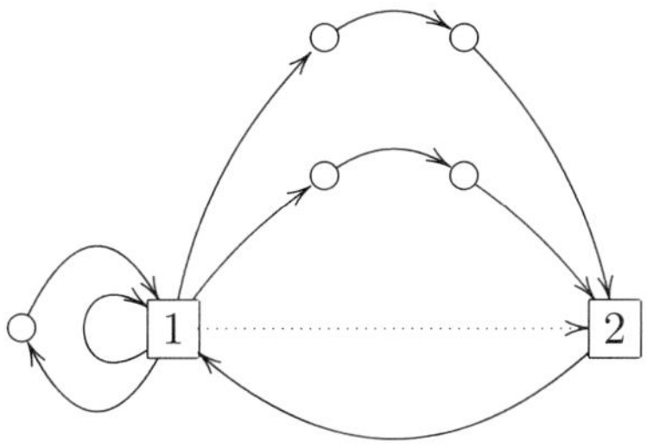

where the constant term route is shown as a dotted arrow and the power term routes are shown as solid arrows. The nonnegative polynomial matrix A and the graph $A^\flat$ are essentially identical ways of presenting the same data, and we will consider them as being the same.

We will explain two equivalent methods for constructing shift spaces from $A^\flat$. The first space $(P(A), \sigma(A))$ will use the *path space construction* which generalizes the well known edge path construction explained in [LM]. This will useful in showing how certain elementary row and column operations on the matrix $I - A$ give rise to topological conjugacies of shift spaces. The second space $(X(A), \sigma(A))$ will come from a zero-one matrix $A^\#$ of finite size. One of its uses is to show $(P(A), \sigma(A))$ is actually a subshift of finite type.

The Path Space Construction

Consider the set P of all sequences $E = \{E_n\} = \{(r_n, t_n)\}$ where $-\infty < n < \infty$ satisfying the conditions

(1) Each r_n is a route of $A^\flat$, and the end vertex of r_n is the start vertex of r_{n+1}.

(2) Each t_n is an integer and $t_{n+1} = t_n + |r_n|$ where $|r_n|$ is zero if r_n is a constant term route and is the number of arcs in r_n if r_n is a power term route.

The intuitive idea is that E is an infinite trip through $A^\flat$ where at time t_n the voyager is at the starting vertex of the route r_n and is about to traverse r_n. Let Ω denote the (infinite) alphabet consisting of all the routes of $A^\flat$ together with the integers. We endow P with the topology it inherits as a subset of the infinite product Ω^Z. Say $E = \{(r_n, t_n)\}$ is *equivalent* to $E' = \{(r'_n, t'_n)\}$ and write $E \sim E'$ provided there is some integer m so that $r_{n+m} = r'_n$ and $t_{n+m} = t'_n$ for all n. Use the quotient topology to define

$$P(A) = P \text{ modulo } \sim \tag{4.2}$$

We remark that $P(A)$ is compact. To see this, let L be the length of the longest route in $A^\flat$. Let P_L denote the compact subspace of P consisting of those E satisfying $0 \leq t_0 \leq L$. Then the quotient map from P_L to $P(A)$ is still onto, because the NZC implies that any E must have $0 \leq t_n \leq L$ for some n. Next define a shift map $\sigma : P \to P$ by the equation

$$\sigma(E)_n = (r_n, t_n - 1) \tag{4.3}$$

This respects the equivalence relation $\sim$ and induces a continuous shift map

$$\sigma(A) : P(A) \to P(A) \tag{4.4}$$

The #-Construction

We will now construct a zero-one matrix $A^{\#}$ with finite support from the graph $A^{\flat}$, and by definition, we will let

$$(X(A), \sigma(A)) = \{X_{A^{\#}}, \sigma_{A^{\#}}\} \tag{4.5}$$

Special Case $A(0) = 0$

This is just the edge path construction for the graph $A^{\flat}$. Namely, the set of states $S^{\#}$ is the set of arcs between primary and/or secondary vertices in $A^{\flat}$, and $A^{\#}(\alpha, \beta) = 1$ iff the end vertex of α is the start vertex of β.

General Case

The first step is to define the set $S^{\#}$ of states of $A^{\#}$. The states will be quintuples

$$\alpha = (i, \alpha', j, \alpha'', k) \tag{4.6}$$

where

(1) Each of i, j, and k is a primary vertex,
(2) If $i \neq j$, then α' is a connected path of constant term routes from i to j and α'' is a subarc of a power term route from j to k.
(3) If $i = j$, then there are no paths of constant term routes from i to j because of NZC. So we let α' be the symbol ϕ, and we require α'' to be a subarc of a power term route from j to k.

When $i = j$, the state α is really just the triple (j, α'', k). So, for example, if $A_{jk} = t$, then α is identified with the power term arc from j to k. But we also write α as a quintuple as in (4.6) to preserve uniformity of notation. There are no states consisting only of a constant term path from one vertex to another. NZC implies that the number of paths consisting only of constant term routes is finite. Therefore $S^{\#}$ is a finite set, and $(X_{A^{\#}}, \sigma_{A^{\#}})$ will be a subshift of finite type.

The next step is to define the transition function

$$A^{\#} : S^{\#} \times S^{\#} \to \{0, 1\} \tag{4.7}$$

Let $\alpha = (i, \alpha', j, \alpha'', k)$ and $\beta = (p, \beta', q, \beta'', r)$. Then

$$\begin{array}{ll} A^{\#}(\alpha,\beta)=1 & \text{when the conditions in Case 1} \\ & \text{or Case 2 below hold} \\ A^{\#}(\alpha,\beta)=0 & \text{otherwise} \end{array}$$

Case 1. We have $i=p$, $j=q$, $k=r$, and $\alpha'=\beta'$. Both α'' and β'' are subarcs on the same power term route from j to k and the ending vertex of α'' is the starting vertex of β''.

Case 2. The subarc α'' is the last one along a power term route from j to k and therefore the ending vertex of α'' is k. We require $k=p$, and we require that the subarc β'' is the first one along a power term route from q to r. In particular, the beginning vertex of β'' is q.

The final step is to make $A^{\#}$ into an $n^{\#}\times n^{\#}$ matrix where $n^{\#}$ is the number of states. We do this by

(4.8) choosing an ordering of the set $S^{\#}$

Two different choices of orderings of $S^{\#}$ will produce matrices $A_1^{\#}$ and $A_2^{\#}$ related by the equation

$$A_2^{\#}=Q^{-1}A_1^{\#}Q \tag{4.9}$$

where Q is a permutation matrix. In particular, Q induces a conjugacy from $(X_{A_1^{\#}},\sigma_{A_1^{\#}})$ to $(X_{A_2^{\#}},\sigma_{A_2^{\#}})$.

In the case where the constant term matrix $A(0)=0$, there are no constant term routes and the matrix $A^{\#}$ is the same as the one constructed in [KRW2, KRW3]. If A is a matrix over Z^{+}, then $\{tA\}^{\flat}$ is the graph associated to the matrix A and $\{tA\}^{\#}$ is precisely the zero-one matrix describing the edge path presentation matrix $A^{'}$ of the subshift of finite type associated to A. In particular, we have

$$(X(tA),\sigma(tA))=\{X_{A'},\sigma_{A'}\} \tag{4.10}$$

See [LM].

Example 4.11 Consider the matrix

$$A=\begin{pmatrix} t & 1 \\ t & 0 \end{pmatrix}$$

The graph $A^{\flat}$ is

There are three states:

$\alpha_1 =$

$\alpha_2 =$

$\alpha_3 =$

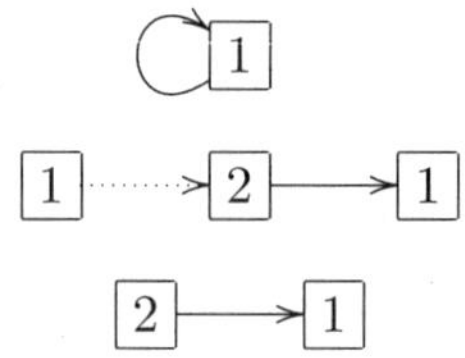

The matrix $A^{\#}$ is

$$\begin{pmatrix} 1 & 1 & 0 \\ 1 & 1 & 0 \\ 1 & 1 & 0 \end{pmatrix}$$

Hence

$$(X(A), \sigma(A)) = \{X_{A^{\#}}, \sigma_{A^{\#}}\} = \text{ the full Bernoulli 2-shift}$$

As discussed in the Appendix to [BW], we don't require each row and each column of a matrix M to have a nonzero entry in forming $\{X_M, \sigma_M\}$.

Example 4.12 Consider the matrix

$$A = \begin{pmatrix} t & t \\ 1 & 0 \end{pmatrix}$$

The graph $A^\flat$ is

There are four states:

$\alpha_1 =$

$\alpha_2 =$

$\alpha_3 =$

$\alpha_4 =$

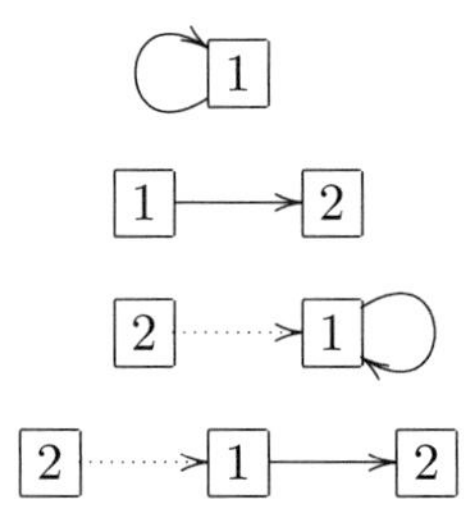

The matrix $A^{\#}$ is

$$\begin{pmatrix} 1 & 1 & 0 & 0 \\ 0 & 0 & 1 & 1 \\ 1 & 1 & 0 & 0 \\ 0 & 0 & 1 & 1 \end{pmatrix}$$

which is just the matrix for the edge path presentation, as discussed in [LM], of the subshift of finite type associated to the graph

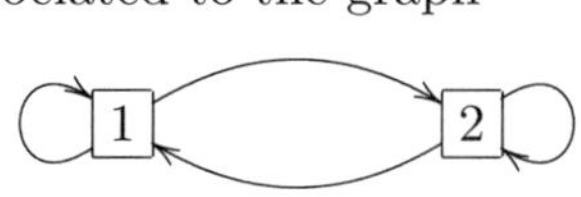

Hence, we again have

$$(X(A), \sigma(A)) = \{X_{A^{\#}}, \sigma_{A^{\#}}\} = \text{ the full Bernoulli 2-shift}$$

Equivalency of the Path Construction and the #-Constuction

THEOREM 4.13. *There is a topological conjugacy* $\Phi : (P(A), \sigma(A)) \to (X(A), \sigma(A))$.

We briefly show how Φ is constructed. Let $E = \{(r_n, t_n)\}$ represent an infinite path in $P(A)$. We want to find an infinite allowable sequence $\Phi(E) = \{\Phi(E)_n\}$ of states in $S^{\#}$. The indices n of the states $\Phi(E)_n$ run through all the integers. This set can which can be written as the union of intervals $[t_p, t_{p+1}, \ldots, t_{p+q+1}]$ where $t_{p-1} < t_p = \ldots = t_{p+q} < t_{p+q+1}$. The intervals overlap only at their endpoints. Let i be the start vertex of r_p, j be the start vertex of r_{p+q}, and k be the final vertex of r_{p+q}. Let α' be the path of constant term routes leading from i to j which is the concatenation of $r_p, \ldots, r_{p+q-1}$. The power term route r_{p+q} is the concatenation of subarcs $\alpha''_1, \ldots, \alpha''_\ell$ where $t_{p+q+1} = t_{p+q} + \ell = t_p + \ell$. Define

$$\Phi(E)_n = (i, \alpha', j, \alpha''_{n-t_p+1}, k) \quad \text{for } t_p \leq n < t_{p+q+1} \tag{4.14}$$

Entropy and the Zeta Function

Polynomial matrices are very helpful in computing basic invariants of subshifts such as entropy and the Artin-Mazur zeta function. The proof of (1.7) in [BGMY] shows that

$$\begin{aligned} det(tI - A^{\#}) &= t^v det(I - A(t^{-1})) \\ det(I - tA^{\#}) &= det(I - A) \end{aligned} \tag{4.15}$$

where v denotes the number of vertices in $A^{\#}$ and where $A(t^{-1})$ is obtained from A by substituting t^{-1} for t. Consequently, we have

$$\text{Entropy of } (X(A), \sigma(A)) = \text{largest root of } t^v det(I - A(t^{-1})) \tag{4.16}$$

$$\zeta_A(t) = \frac{1}{det(I - A)} \tag{4.17}$$

The Dimension Group

Let n be a positive integer. Let F^n be the standard free, left $Z[t]$-module of rank n. Let F^∞ denote the free, left $Z[t]$-module which is the direct sum of a countable number of copies of $Z[t]$. Define

$$G(A) = Coker(I - A) = F^\infty / Image(I - A) \tag{4.18}$$

This is finitely generated because A has only finitely many non-zero entries. Let F^∞_+ denote the subset of F^∞ consisting of elements that have all coordinates non-negative. Define

$$G^+(A) = \text{Image of } F^\infty_+ \text{ in } G(A) \tag{4.19}$$

If A is a nonnegative integral matrix, then there is an isomorphism

$$(G(tA), G^+(tA), s(tA)) \equiv (G_A, G_A^+, s_A) \tag{4.20}$$

THEOREM 4.21. *Assume A is a nonnegative polynomial matrix satisfying NZC. The action of t on $G(A)$ is an isomorphism which takes $G^+(A)$ bijectively to itself. In particular, $G(A)$ is a module over the ring $Z[t,t^{-1}]$ of Laurent polynomials and $s(A) = t^{-1}$ is an isomorphism of the pair $(G(A), G^+(A))$. Moreover, there is an isomorphism*

$$((G(A), G^+(A), s(A)) \equiv (G_{A\#}, G_{A\#}^+, s_{A\#})$$

of $Z[t,t^{-1}]$-modules.

See [BW] and (2.12) of [KRW2]. When $A(0) = 0$, we can write $A = tB$, and it is clear from the definition $G(A) = coker(I - tB)$ that the action of t is invertible.

Small representations

The results above show there is great ecomony in using polynomial matrices to represent subshifts of finite type. Let A be a primitive, nonnegative integral matrix. A natural and still open question is

(4.22) What is the minimal size of a nonnegative polynomial matrix M such that $(X(M), \sigma(M))$ is topologically conjugate to (X_A, σ_A) ?

Some recent progress has been made in this area by Boyle and Lind in [BL]. They study M for which $M^\#$ has a spectrum closely related to the spectrum of A. Two of their results are (4.23) and (4.24) below. It turns out that we may assume $M(0) = 0$ without loss of generality.

As in [BL] define a *list* to be a collection of complex numbers where the multiplicity matters but the order does not. Thus $\{1,2,2,3\} = \{3,2,1,2\} \neq \{1,2,3\}$. The list Λ is *contained* in the list Λ', written $\Lambda \subset \Lambda'$, provided the multiplicity of each λ in Λ is less than or equal to its multiplicity in Λ'. The *spectral radius* of Λ is $\rho(\Lambda) = \max_{\lambda\epsilon\Lambda}|\lambda|$. A list Λ is *Perron* if there is a positive real number λ in Λ of multiplicity one such that $\lambda > |\mu|$ for every other μ in Λ. In particular, $\rho(\Lambda) = \lambda$. The list Λ is *Z-algebraic* provided the polynomial $f_\Lambda(t) = \prod_{\lambda\epsilon\Lambda}(t-\lambda)$ has integer coefficients. If A is a real matrix, let $sp(A)$ denote the list of complex numbers which are eigenvalues of A and let $sp^\times(A)$ be the list of non-zero eigenvalues of A. Then the spectral radius of A is just the spectral radius of $sp(A)$.

THEOREM 4.23. *Let Λ be a Z-algebraic Perron list of non-zero complex numbers. Then there is a nonnegative polynomial matrix M of size at most 2×2 such that $M^\#$ is primitive, $\rho(\Lambda) = \rho(M^\#)$, and $\Lambda \subset sp^\times(M^\#)$.*

Let M be a nonnegative polynomial matrix. Define $g(M)$ to be smallest cardinality of a set of generators of $G(M)$ considered as a $Z[t,t^{-1}]$-module. Let A be a nonnegative integral matrix of finite support. If $(X(M), \sigma(M))$ is topologically conjugate to (X_A, σ_A), then we would have an isomorphism

$$(G(M), G^+(M), s(M)) = (G(tA), G^+(tA), s(tA))$$

More generally, suppose M is an $r \times r$ polynomial matrix so that there is a surjection of $Z[t, t^{-1}]$-modules from $G(M)$ to $G(tA)$. Then we would have

$$g(tA) \leq g(M) \leq r$$

So $g(tA)$ is a lower bound on the size of M.

THEOREM 4.24. *Let A be a primitive, nonnegative integral matrix. Then there is a nonnegative polynomial matrix M of size at most $g(tA)+1$ such that $\rho(M^{\#}) = \rho(A)$, $M^{\#}$ is primitive, and there is a surjection of $Z[t, t^{-1}]$-modules from $G(M)$ to $G(tA)$.*

The actual results in [BL] are more general than those stated above and work with matrices over subrings S of the real numbers and polynomials with coefficients in S. Also, instead of using the #-construction, Boyle and Lind use a variation of it called the $\natural$-construction for matrices with entries in $S[t]$. When $S = Z$ and $M = tP$ where P is a matrix with nonnegative integral polynomial entries, then $P^{\natural}$ is strong shift equivalent over Z^+ to $M^{\#}$.

Simple Automorphisms

For future reference, we now generalize the usual definition [N,W4] of simple automorphisms of a subshift of finite type. An *elementary simple automorphism* of $(X(A), \sigma(A))$ is one coming from an automorphism of the graph $A^{\flat}$ which fixes the primary vertices. A *simple automorphism* of $(X(A), \sigma(A))$ is one of the form $\alpha\Theta\alpha^{-1}$ where $\alpha : (X(A), \sigma(A)) \to (X(B), \sigma(B))$ is a topological conjugacy and Θ is an elementay simple automorphism of $(X(B), \sigma(B))$. We let $Simp(\sigma(A))$ denote the subgroup of $Aut(\sigma(A))$ generated by simple automorphisms. It is a fact that this definition of simple automorphisms gives the same subgroup of simple automorphisms as does the definition in Section 3.

5. Nonnegative Row and Column Operations

In this section we discuss how nonnegative row and column operations on matrices give rise to conjugacies between subshifts of finite type defined by matrices satisfying NZC. This is closely related to the well known technique of state splitting and merging discussed in [LM].

Let $A = \{A_{ij}\}$ be a nonnegative polynomial matrix with finite support satisfying NZC. Fix an entry A_{kl} of A where $k \neq l$, and let b denote a polynomial in $Z[t]$. Let $T_{kl}(b)$ denote the matrix which has the entry b in the kth row and l^{th} column and zeros elsewhere. Let I denote the infinite identity matrix. Define

$$E_{kl}(b) = I + T_{kl}(b) \tag{5.1}$$

$E_{kl}(b)$ will be called a *nonnegative shear* if b lies in $Z^+[t]$. Define a new matrix of finite support B by one of the equations

$$
\begin{array}{rcl}
I - B & = & E_{kl}(b)(I - A) \\
I - B & = & (I - A)E_{kl}(b)
\end{array}
\tag{5.2}
$$

We will say that $E_{ij}(b)$ *goes from* $I - A$ *to* $I - B$.

If $a(t) = \sum_r a_r t^r$ and $b(t) = \sum_r b_r t^r$ are polynomials, we define $a \leq b$ iff $a_r \leq b_r$ for each r.

THEOREM 5.3. *Assume* $0 \leq b \leq A_{kl}$. *Then* B *is nonnegative and satisfies NZC. Corresponding to the first and second equations in (5.2) respectively there are conjugacies*

$$L_{kl}(b) : (X(A), \sigma(A)) \to (X(B), \sigma(B))$$
$$R_{kl}(b) : (X(A), \sigma(A)) \to (X(B), \sigma(B))$$

$L_{kl}(b)$ *and* $R_{kl}(b)$ *are well defined up to multiplication on the right by elements in* $Simp(\sigma(A))$ *and multiplication on the left by elements in* $Simp(\sigma(B))$.

There is also a version of (5.3) over $ZO[t]$, the set of *zero-one polynomials*. By definition, $ZO[t]$ consists of those polynomials $a(t) = \sum_r a_r t^r$ where each coefficient a_r is equal to zero or one. The sum and product of two zero-one polynomials is not always a zero-one polynomial. Nevertheless, there are still many cases when the equations (5.2) do hold over $ZO[t]$. This means that A has zero-one polynomials as entries, b is a zero-one polynomial satisfying $b \leq A_{kl}$, and all the additions and multiplications occurring on the right hand side of (5.2) produce zero-one polynomials. Finally, we require the matrix B defined in (5.2) to have zero-one polynomial entries. For example, if $(P, Q) : M \to N$ is a strong shift equivalence in the category of zero-one matrices, then the PSSE equations below are examples where (5.2) holds over $ZO[t]$. In the category of matrices over $ZO[t]$, the conjugacies $L_{kl}(b)$ and $R_{kl}(b)$ are well defined without any simple automorphism ambiguity.

Indication of Proof

We will discuss $L = L_{kl}(b)$ in the context of $(P(A), \sigma(A))$. For simplicity, let $b = t^p$ be a single route from the primary vertex k to the primary vertex l of length $p \geq 0$. The matrix B (i.e., the graph $B^\flat$) is obtained from the matrix A by first deleting the route b between k and l and then inserting a route r' of length $|r'| = p + |r|$ from k to q for each route r starting at l and ending at q. Every infinite trip through A corresponds uniquely to a concatenation of routes in B and this determines L. More precisely, let $E = \{(r_n, t_n)\}$ be in $P(A)$. Consider a part of E like

$$\ldots, (r_n, t_n), (r_{n+1}, t_{n+1}), (r_{n+2}, t_{n+2}), \ldots$$

where $r_n = b$ and $r_{n+1} = r$ as above. Delete (r_{n+1}, t_{n+1}) and replace (r_n, t_n) with (r', t_n). Note that $t_{n+2} = t_n + |r'|$. The result is just a subsequence E' of E where some of the items have been changed. Now renumber this subsequence in an increasing fashion so that it becomes a sequence $E' = \{E'_n\}$ where n runs through *all* the integers. The equivalence class modulo $\sim$ of E' does not depend on this renumbering, and we define $L(E) = E'$.

Classification

We now summarize the discussion of conjugacy and eventual conjugacy [LM] in terms of row and column operations on polynomial matrices.

THEOREM 5.4. *Let A and B be nonnegative polynomial matrices satisfying NZC. Assume the matrices $A^{\#}$ and $B^{\#}$ are strong shift equivalent to primitive matrices over Z^+. Then the following conditions are equivalent:*
(G) There is an isomorphism between the $Z[t,t^{-1}]$-modules $G(A)$ and $G(B)$
(G^+) There is an isomorphism between the dimension module triples $(G(A), G^+(A), s(A))$ and $(G(B), G^+(B), s(B))$
(E) There is a sequence of row and column operations over $Z[t]$. connecting $I - A$ and $I - B$

THEOREM 5.5. *Let A and B be nonnegative polynomial matrices satisfying NZC. There is a conjugacy between $(X(A), \sigma(A))$ and $(X(B), \sigma(B))$ iff there is a path of nonnegative row and column operations over $Z^+[t]$ connecting $I - A$ and $I - B$.*

THEOREM 5.6. *Let A and B be nonnegative polynomial matrices satisfying NZC. Assume the matrices $A^{\#}$ and $B^{\#}$ are strong shift equivalent to primitive matrices over Z^+. There is an eventual conjugacy between $(X(A), \sigma(A))$ and $(X(B), \sigma(B))$ iff there is a path of row and column operations over $Z[t]$ connecting $I - A$ and $I - B$.*

Let A and B be nonnegative integral matrices. Recall from Section 2 that (X_A, σ_A) and (X_B, σ_B) are conjugate iff A and B are strong shift equivalent over Z^+, (X_A, σ_A) and (X_B, σ_B) are eventually conjugate iff the matrices A and B are shift equivalent over Z^+, and that primitive matrices A and B are shift equivalent over Z^+ iff they are strong shift equivalent over Z. So parallel to the discussion in Section 2 we have the following results.

THEOREM 5.7. *The matrices A and B are strong shift equivalent over Z^+ iff there is a path of nonnegative row and column operations over $Z^+[t]$ connecting $I - tA$ and $I - tB$.*

THEOREM 5.8. *Assume A and B are primitive matrices. The matrices A and B are strong shift equivalent over Z iff there is a path of row and column operations over $Z[t]$ connecting $I - tA$ and $I - tB$.*

Here is an example of (5.5) for the matrices

$$A = \begin{pmatrix} t & 1 \\ t & 0 \end{pmatrix} \quad \text{and} \quad B = \begin{pmatrix} t & t \\ 1 & 0 \end{pmatrix}$$

Both give the shifts conjugate to (X_2, σ_2) which comes from the matrix

$$C = \begin{pmatrix} 2t & 0 \\ 0 & 0 \end{pmatrix}$$

$I - C$ is obtained from $I - A$ by first multiplying on the left by $E_{12}(1)$ and then on the right by $E_{21}(t)$. $I - C$ is obtained from $I - B$ by first multiplying on the left by $E_{12}(t)$ and then on the right by $E_{21}(1)$.

Flow Equivalence

By considering nonnegative row and column operations over Z^+ instead of $Z^+[t]$, the methods above can be modified slightly to be useful in the flow equivalence classification of the suspension of subshifts of finite type and in the discussion of the path components of the topological group of flow equivalences, i.e., the mapping class group in this context. See [Ba] and [Bo4]. In particular, these methods are crucial to the proof that the Bowen-Franks representation of the mapping class group is surjective for the suspension of a nontrivial subshift of finite type.

6. The Δ-strategy

The exact homotopy sequence of the pair $(SSE(\Lambda), SSE(\Lambda^+))$ leads to a strategy for producing counterexamples to the Strong Shift Equivalence Problem which was observed independently by Kim-Roush and by the author.

Let Λ be a ring containing 1 and with a set of *nonnegative* elements Λ^+ which contains 0 and 1 and which is closed under addition and multiplication. We will assume Λ^+ satisfies the condition that if a and b are in Λ^+ and $a+b=0$ or $ab=0$, then either $a=0$ or $b=0$. Typical examples of Λ are the integers Z, a subring of the real numbers, a ring of polynomials in commuting or noncommuting variables with integer coefficients, a ring of Laurent polynomials with integer coefficients, and the integral group ring $Z[G]$ of a group G.

Let S^+ denote a union of components in $SSE(\Lambda^+)$. For example S^+ could be $SSE_m(\Lambda^+)$, which consists of those components of $SSE(\Lambda^+)$ containing vertices A satisfying

$$Trace(A) = \cdots = Trace(A^m) = 0$$

Let $\pi_1(SSE(\Lambda), S^+)$ denote homotopy classes of paths in $SSE(\Lambda)$ with endpoints in S^+. We would like to find a function

$$\Delta : \pi_1(SSE(\Lambda), S^+) \to G \tag{6.1}$$

where G is, say, an abelian group such that Δ satisfies the properties

$$\begin{aligned} &\Delta(\alpha * \beta) = \Delta(\alpha) + \Delta(\beta) \\ &\Delta(\alpha) = 0 \text{ whenever } \alpha \text{ lies in } S^+ \end{aligned} \tag{6.2}$$

Let A and B be vertices in S^+ and choose a path α from A to B in $SSE(\Lambda)$. If β is another path from A to B we have

$$\Delta(\alpha) = \Delta(\beta) + \Delta(\alpha * \beta^{-1})$$

Consequently, there is an invariant

$$\Delta(A, B) = \Delta(\beta) \quad \text{in} \quad G \bmod \Delta(\pi_1(SSE(\Lambda), A)) \tag{6.3}$$

which vanishes if there is a path from A to B in S^+. In particular, a counterexample to the Strong Shift Equivalence Problem can be obtained by finding a function

Δ together with matrices A and B such that

$$\text{(6.4)} \qquad \begin{array}{l} \Delta(A,B) = \Delta(\beta) \neq 0 \text{ for some } \beta \\ \Delta(\alpha) = 0 \text{ whenever } \alpha \text{ is in } \pi_1(SSE(\Lambda), A) \end{array}$$

7. An SSE invariant in K_2 of the dual numbers

Two invertible matrices X and Y over a ring R determine the same element in the algebraic K-theory group $K_1(R)$ iff $EXF = Y$ where E and F are products of elementary matrices $e_{ij}(r)$. By definition, $e_{ij}(r)$ is the invertible matrix which is the identity I on the diagonal, has the element r of R in the ith row and jth column where $i \neq j$, and is zero elsewhere. Roughly speaking, the abelian group $K_2(R)$ is a measure of the different ways to get from X to Y using such row and column operations. In particular, any product of elementary matrices

$$e_{i_1j_1}(r_1)e_{i_2j_2}(r_2)\cdots e_{i_pj_p}(r_p) = I$$

gives rise to an element of $K_2(R)$. See [Mi, R].

Here are the precise definitions of $K_1(R)$ and $K_2(R)$. Let $E(R)$ denote the subgroup of the general linear group $Gl(R)$ generated by the elementary matrices $e_{ij}(r)$. The Whitehead Lemma says that $E(R)$ is the commutator subgroup of $Gl(R)$. Define the abelian group $K_1(R)$ by the equation

$$\text{(7.1)} \qquad K_1(R) = Gl(R)/[Gl(R), Gl(R)] = Gl(R)/E(R)$$

If R is a commutative, ring then the determinant function $det(A)$ of a matrix A defines a homomorphism

$$det : K_1(R) \to R^*$$

which is actually an isomorphism for many rings. This is the case, for example, if R is a principal ideal domain, a field, or the ring of integers in a number field. To obtain K_2, first define the *Steinberg group* $St(R)$ by generators $x_{ij}(r)$ and relations

(ST1) $x_{ij}(r+s) = x_{ij}(r)x_{ij}(s)$
(ST2) $[x_{ij}(r), x_{kl}(s)] = 1$ if $i \neq l$ and $j \neq k$
(ST3) $[x_{ij}(r), x_{jk}(s)] = x_{ik}(rs)$ if i, j, k are distinct

These relations are satisfied by the corresponding matrices $e_{ij}(r)$, and so the function $\phi(x_{ij}(r)) = e_{ij}(r)$ defines a homomorphism

$$\phi : St(R) \to E(R)$$

By definition

$$\text{(7.2)} \qquad K_2(R) = \text{Kernel of } \phi$$

The resulting exact sequence

$$0 \to K_2(R) \to St(R) \to E(R) \to 1$$

is the universal central extension of $E(R)$. This definition of K_2, due to John Milnor, is related to Robert Steinberg's work on universal covering groups of algebraic groups. K_2 is also closely connected to reciprocity laws in algebraic number theory.

Consider the truncated polynomial ring $\Lambda[t]/(t^{m+1})$, which is called the *dual numbers* when $m = 1$. Let $K_2(\Lambda[t]/(t^{m+1}), (t))$ be the cokernel of the split injection $K_2(\Lambda) \to K_2(\Lambda[t]/(t^{m+1}))$. For example , van der Kallen [vdK] showed that

$$K_2(Z[t]/(t^2), (t)) \simeq Z/2Z$$

and

$$K_2(\Lambda[t]/(t^2), (t)) \simeq \Omega^1_{\Lambda/Z} = K\ddot{a}hler\ differentials$$

if Λ is a commutative ring containing $1/2$. Further information on these relative K-groups may be found in [vdK,Bl,Ke,Lo,MS,St].

THEOREM 7.3. *The procedure in Step 1, Step 2, and Step 3 below defines a function*

$$\Phi_{2m} : \pi_1(SSE(\Lambda), SSE_{2m}(\Lambda^+)) \to K_2(\Lambda[t]/(t^{m+1}), (t))$$

satisfying (6.2). Moreover, assume $\Lambda = Z$ and consider a path $(R, S) : A \to B$ in $SSE(Z)$ where A and B are nonnegative matrices with $Trace(A) = Trace(A^2) = Trace(B) = Trace(B^2) = 0$. Then $\Phi_2(R, S) = sgc_2(R, S)$ in $Z/2Z$ where

$$sgc_2(R,S) = \sum_{i<j,k>l} R_{ik}S_{ki}R_{jl}S_{lj} + \sum_{i<j,k\geq l} R_{ik}S_{kj}R_{jl}S_{li} + \sum_{i,k} \frac{R_{ik}(R_{ik}-1)}{2} S^2_{ki}$$

The function Φ_{2m} was constructed in [W8] where an algorithm for evaluating $\Phi_2(R, S)$ when $\Lambda = Z$ was given using van der Kallen's paper [vdK]. Machine computations of Φ_2 also detected the counterexamples to the Strong Shift Equivalence Problem given in Section 8 below and [KR6]. The *sign-gyration-compatibility* cohomology class sgc_2 was used by Kim and Roush in [KR6] to find the first counterexamples to the Shift Equivalence Problem. See [W7,KRW4] for an exposition of this method. Further numerical evidence raised the question whether $\Phi_2 = sgc_2$, and this was subsequently shown to be true by Kim-Roush in the Appendix to [W8].

The machinery of algebraic K-theory may be useful in computing the indeterminacy which occurs in (6.3) and (6.4) in the strategy for finding counterexamples to the Strong Shift Equivalence Problem for other rings Λ. Assume Λ is a commutative ring, so that $det(I - tA)$ makes sense. It was shown in [W8] that $det(I - tA) = 1 \mod t^{m+1}$ whenever $Tr(A) = \ldots = Tr(A^m) = 0$ and $A \geq 0$ and that the image of Φ_{2m} on $\pi_1(SSE(\Lambda), A)$ is contained in the image of the resulting homomorphism

$$K_2(\Lambda[t][\frac{1}{det(I-tA)}]) \to K_2(\Lambda[t]/(t^{m+1})) \to K_2(\Lambda[t]/(t^{m+1}), (t))$$

If Λ is a Noetherian ring, there is the localization exact sequence

$$\ldots K_2(\Lambda[t]) \to K_2(\Lambda[t][\frac{1}{det(I-tA)}]) \to K_1(\Lambda[t]/(det(I-tA))) \to K_1(\Lambda[t]) \ldots$$

See [Lo, p.357] or [R, p.294]. If Λ is also regular [R, p.110], then $K_n(\Lambda[t]) = K_n(\Lambda)$ for all n [R, p.295], the localization sequence is split exact, and we have isomorphisms

$$K_2(\Lambda[t][\frac{1}{det(I-tA)}])/K_2(\Lambda) \simeq K_1(\Lambda[t]/(det(I-tA)))$$

This discussion gives an "explanation" of what is happening when we show that $sgc_2 = \Phi_2 = 0$ on $\pi_1(SSE(Z), A) = Aut(s_A) = Z[A]^* = K_1(Z[A])$ for the particular A appearing in Section 8. It also shows that computing generators for $K_1(\Lambda[t]/(det(I - tA)))$ is key for examining examples over other rings Λ. The Laurent polynomial ring $\Lambda = Z[t_1, t_1^{-1}, \ldots, t_n, t_n^{-1}]$ comes up in studying Markov chains [MT1-MT3,T].

Here is an outline of the construction of Φ_2. On the one hand, there are the strong shift equivalence equations and the strong shift equivalence spaces of Section 3. On the other hand, the algebraic K-theory groups K_1 and K_2 arise from row and column operations on matrices. The basic idea is to use the *polynomial strong shift equations* of [BW, W8] which convert an elementary strong shift equivalence $(P, Q) : M \to N$ into a sequence of row and column matrix operations leading from $I - tM$ to $I - tN$. This is analogous to one-parameter families of functions which pseudo-isotopy theory [C,HW] relates to K_2.

PSSE Equations

$$\begin{pmatrix} I - tPQ & 0 \\ -tQ & I \end{pmatrix} \begin{pmatrix} I & 0 \\ tQ & I \end{pmatrix} = \begin{pmatrix} I - tPQ & 0 \\ 0 & I \end{pmatrix}$$

$$\begin{pmatrix} I & P \\ 0 & I \end{pmatrix} \begin{pmatrix} I & -P \\ -tQ & I \end{pmatrix} = \begin{pmatrix} I - tPQ & 0 \\ -tQ & I \end{pmatrix}$$

$$\begin{pmatrix} I & -P \\ -tQ & I \end{pmatrix} \begin{pmatrix} I & P \\ 0 & I \end{pmatrix} = \begin{pmatrix} I & 0 \\ -tQ & I - tQP \end{pmatrix}$$

$$\begin{pmatrix} I & 0 \\ tQ & I \end{pmatrix} \begin{pmatrix} I & 0 \\ -tQ & I - tQP \end{pmatrix} = \begin{pmatrix} I & 0 \\ 0 & I - tQP \end{pmatrix}$$

The strong shift equivalence equations led the author to the polynomial strong shift equations in the early 1980's , and this suggested that it might be fruitful to explore analogies between algebraic K-theory and symbolic dynamics. The result was the SSE spaces and sgc_k cohomology classes. Then in the last few years the polynomial matrix viewpoint reappeared and has been used in [Bo2,BL,BW,KRW2,KRW3, KOR,W8]. Part of this effort was Mike Boyle's independent rediscovery of the polynomial strong shift equations in the 1990's.

Step 1. Let $(P, Q) : M \to N$ be a strong shift equivalence over Λ. Consider the products of elementary matrices

$$W_L(P,Q) = \begin{pmatrix} I & 0 \\ tQ & I \end{pmatrix} \begin{pmatrix} I & -P \\ 0 & I \end{pmatrix}$$

$$W_R(P,Q) = \begin{pmatrix} I & 0 \\ -tQ & I \end{pmatrix} \begin{pmatrix} I & P \\ 0 & I \end{pmatrix}$$

$$E_L(P,Q) = W_L(I,N)^{-1} W_L(P,Q)$$

$$E_R(P,Q) = W_R(P,Q) W_R(I,N)^{-1}$$

We have the matrix equation

$$(7.4) \qquad E_L(P,Q) \begin{pmatrix} I - tM & 0 \\ 0 & I \end{pmatrix} E_R(P,Q) = \begin{pmatrix} I - tN & 0 \\ 0 & I \end{pmatrix}$$

Step 2. Let M be a matrix over Λ^+ such that $Tr(M) = 0$. Then each diagonal entry M_{ii} is zero. Consider the products of elementary matrices

$$L_M = \prod_{j=1}^{s-1} \prod_{i=j+1}^{s} e_{ij}(tM_{ij})$$

$$R_M = \prod_{i=1}^{s-1} \prod_{j=i+1}^{s} e_{ij}(tM_{ij})$$

We have the matrix equation

$$(7.5) \qquad L_M(I - tM)R_M = I - t^2 M'$$

for some matrix M' over the polynomial ring $\Lambda[t]$.

Step 3. Let A and B be matrices over Λ^+ with $Tr(A) = Tr(B) = 0$, and let γ be a path in $SSE(\Lambda)$ connecting A and B. Apply Step 1 to each arc $(P, Q) : M \to N$ in γ to get products of elementary matrices $E_L(\gamma)$ and $E_R(\gamma)$ satisfying the matrix equation

$$(7.6) \qquad E_L(\gamma)(I - tA)E_R(\gamma) = I - tB$$

Step 2 yields

$$L_B E_L(\gamma) L_A^{-1}(I - t^2 A') R_A^{-1} E_R(\gamma) R_B = I - t^2 B'$$

Setting $t^2 = 0$ gives the matrix equation

$$(7.7) \qquad L_B E_L(\gamma) L_A^{-1} R_A^{-1} E_R(\gamma) R_B = I$$

in the dual numbers $\Lambda[t]/(t^2)$, and this defines $\Phi_2(\gamma)$ in $K_2(\Lambda[t]/(t^2))$.

Then one proves additivity, homotopy invariance with end points fixed, and vanishing whenever γ lies in $SSE_2(\Lambda^+)$. Homotopy invariance under deformation across a triangle in $SSE(\Lambda)$ uses the purely algebraic Exchange Lemma from [HW,W7], which is a consequence of the Steinberg relations and which comes from the geometry of two-parameter families of Morse functions. Vanishing under positivity is proved by starting with the expression

$$\Phi_2(\gamma) = \prod x_{ij}(a)$$

in the Steinberg group $St(\Lambda[t]/(t^2))$ coming from (7.7) and then using the Steinberg relations to reduce it to an expression of the form

$$\prod_{i>j} x_{ij}(b) \prod_{i<j} x_{ij}(c)$$

The reason this can be accomplished is that nonnegativity together with the trace condition $Tr(A) = Tr(A^2) = Tr(B) = Tr(B^2) = 0$ insures expressions of the form $x_{ij}(\alpha)x_{ji}(\beta)$ never obstruct the rearrangement process.

8. A counterexample to the Strong Shift Equivalence Problem for primitive matrices

In this section we give an example of the Δ-strategy for $\Delta = \Phi_2 = sgc_2$. The method for finding candidate matrices A and B comes from [KR5,KR6]. Let M be the 4×4 matrix

$$\begin{pmatrix} 0 & 0 & 1 & 1 \\ 1 & 0 & 0 & 0 \\ 0 & 1 & 0 & 0 \\ 0 & 0 & 1 & 0 \end{pmatrix} \tag{8.1}$$

from [KRW1, 4.1] with characteristic polynomial $t^4 - t - 1$. It satisfies the equation

$$(M - I)(M^4 + M^3 + M^2) = M$$

Let γ be the loop $(E, F) : M \to M$ where

$$E = M - I = \begin{pmatrix} -1 & 0 & 1 & 1 \\ 1 & -1 & 0 & 0 \\ 0 & 1 & -1 & 0 \\ 0 & 0 & 1 & -1 \end{pmatrix} \qquad F = M^4 + M^3 + M^2 = \begin{pmatrix} 2 & 2 & 2 & 1 \\ 1 & 2 & 2 & 1 \\ 1 & 1 & 2 & 1 \\ 1 & 1 & 1 & 1 \end{pmatrix}$$

Direct computation using (7.3) shows $\Delta(\gamma) \neq 0$. Next we want to extend E and F to matrices of the form

$$R = \begin{pmatrix} E & 0 \\ 0 & I \end{pmatrix} \qquad S = \begin{pmatrix} F & X \\ Y & Z \end{pmatrix}$$

so that letting $A = RS$, $B = SR$, and β be the path $(R, S) : A \to B$, we have

(I) A and B are primitive, nonnegative integral matrices with
$det(A) = det(B) = \pm 1$ and
$Trace(A) = Trace(B) = Trace(A^2) = Trace(B^2) = 0$.
(II) $\Delta(\beta) \neq 0$
(III) $\Delta(\alpha) = 0$ for all α in $\pi_1(SSE(Z), A) \simeq Aut(s_A)$.

Trial and error and luck produces

$$X = \begin{pmatrix} 2 & 0 & 0 \\ 2 & 0 & 0 \\ 2 & 0 & 0 \\ 2 & 0 & 0 \end{pmatrix} \quad Y = \begin{pmatrix} 0 & 0 & 0 & 0 \\ 1 & 1 & 2 & 1 \\ 1 & 1 & 2 & 1 \end{pmatrix} \quad Z = \begin{pmatrix} 0 & 0 & 1 \\ 3 & 0 & 0 \\ 0 & 1 & 0 \end{pmatrix}$$

and therefore

$$A = RS = \begin{pmatrix} 0 & 0 & 1 & 1 & 2 & 0 & 0 \\ 1 & 0 & 0 & 0 & 0 & 0 & 0 \\ 0 & 1 & 0 & 0 & 0 & 0 & 0 \\ 0 & 0 & 1 & 0 & 0 & 0 & 0 \\ 0 & 0 & 0 & 0 & 0 & 0 & 1 \\ 1 & 1 & 2 & 1 & 3 & 0 & 0 \\ 1 & 1 & 2 & 1 & 0 & 1 & 0 \end{pmatrix} \quad B = SR = \begin{pmatrix} 0 & 0 & 1 & 1 & 2 & 0 & 0 \\ 1 & 0 & 0 & 0 & 2 & 0 & 0 \\ 0 & 1 & 0 & 0 & 2 & 0 & 0 \\ 0 & 0 & 1 & 0 & 2 & 0 & 0 \\ 0 & 0 & 0 & 0 & 0 & 0 & 1 \\ 0 & 1 & 0 & 0 & 3 & 0 & 0 \\ 0 & 1 & 0 & 0 & 0 & 1 & 0 \end{pmatrix}$$

Direct computation shows Property I holds. Since $\Delta(\gamma) \neq 0$, the formula (7.3) shows that Property II holds for any X and Y and any Z with diagonal entries zero because E was extended by adding on the identity matrix. To verify Property III, first compute the characteristic polynomial for A. It is $t^7 - 6t^4 - 5t^3 - 6t^2 - 37 + 1$. Putting this into the command "buchgenfu()" of PARI gives

$$Z[A]^* = Aut(s_A) = O^* = Z[\lambda]^*$$

in (3.5) because $\{1, \lambda, \lambda^2, \ldots, \lambda^6\}$ is a basis for O. The Dirichlet Unit Theorem shows the rank of O^* is 4, and PARI computes a set of generators for O^* to be

$$R_0 = -1$$

$$R_1 = \lambda$$

$$R_2 = 2\lambda^6 - \lambda^5 + \lambda^4 - 13\lambda^3 - 4\lambda^2 - 13\lambda - 1$$

$$R_3 = \lambda^6 + \lambda^4 - 7\lambda^3 - 6\lambda^2 - 8\lambda - 6$$

$$R_4 = 2\lambda^6 - 4\lambda^5 - 2\lambda^3 + 3\lambda$$

Substitute A for λ in these expressions for each $i = 0, 1, 2, 3, 4$ to get five loops $\alpha_i = (R_i, R_i^{-1}A) : A \to A$ generating $\pi_1(SSE(Z), A) = Aut(s_A)$. $\Delta(\alpha_0) = 0$ because R_0 is diagonal and the diagonal entries of A are zero. Computer computation using the explicit formula in (7.3) shows $\Delta(\alpha_i) = 0$ for each $i = 1, 2, 3, 4$.

9. Inert automorphisms and K_3

Looking beyond the classification of subshifts of finite type, there is the general problem of studying the relation between the spaces $SSE(\Lambda^+)$ and $SSE(\Lambda)$.

Let $\Lambda = Z$. Recall that the group $Inert(\sigma_A)$ of inert automorphisms of $Aut(\sigma_A)$ is the kernel of the dimension group representation (3.8). In view of (3.5), $Aut(s_A)$ is often a finitely generated abelian group [BLR], and consequently much of the complexity of $Aut(\sigma_A)$ typically lies in $Inert(\sigma_A)$. For example, if p is prime, then $Aut(s_p) = Z \oplus \{\pm 1\}$. Moreover, $s_p = \delta(\sigma_p)$ generates the infinite cyclic part of $Aut(s_p)$ and

$$Aut(\sigma_p) = Z \oplus Inert(\sigma_p)$$

From the early work of Hedlund and co-workers [H], the (still open) question

(9.1) Is $Inert(\sigma_p)$ generated by elements of finite order ? By involutions ?

was raised by Frank Rhodes for $p = 2$. During the 1980's this was generalized to the Finite Order Generation Conjecture (FOG)

(9.2) $Inert(\sigma_A)$ is generated by elements of finite order when A is primitive.

The main reason for this conjecture was that historically examples of elements (of finite and infinite order) in $Inert(\sigma_A)$ had been constructed by the classical "marker method" or ingenious extensions thereof, and therefore these elements were always products of elements of finite order. See [BLR]. The first counterexample to FOG was given in [KRW2, KRW3] using the method of nonnegative row and column operations on matrices over $Z^+[t]$ together with the characterization in [BF] of finite order inert actions on periodic points. These methods also give counterexamples to FOG for matrices A where $Trace(A) = Trace(A^2) = 0$ and for higher vanishing trace conditions as well. The question is whether there are "natural" homomorphisms of $Inert(\sigma_A)$ into non-torsion abelian groups which detect these new infinite order elements in $Inert(\sigma_A)$. Here is a proposal. Using (3.6) together with the exact homotopy sequence of the pair $(SSE(Z), SSE(Z^+))$ yields the isomorphism

$$Inert(\sigma_A)/Simp(\sigma_A) = \pi_2(SSE(Z), SSE_2(Z^+); A) \tag{9.3}$$

under the condition that $Trace(A) = Trace(A^2) = 0$. Just as K_2 concerns paths between matrices, K_3 measures deformations between these paths. Each Steinberg relation corresponds to a 2-dimensional cell. The discussion in [W8, Section 7 and Section 8] basically gives a 2-sphere of Steinberg relations and goes halfway towards producing a homomorphism

$$\Theta : Inert(\sigma_A) \to K_3(Z[t]/(t^2), (t)) \tag{9.4}$$

It remains to show this construction is well defined. Maybe higher vanishing trace conditions on A are needed for this. An old result of Soulé says that

$$K_3(Z[t]/(t^2), (t)) \otimes Q = Q$$

A proof using cyclic homology theory can be found in [Lo, 5.4.17 and Chapter 11]. Does the conjectured Θ composed with the cyclic homology Chern character [Lo, Chapter 11] detect infinite order elements in $Inert(\sigma_A)$, at least for certain A ?

There is another connection between strong shift equivalence theory and cyclic homology theory. Namely, the SSE spaces are cyclic spaces [Lo, Chapter 7]. An n-cell of SSE is given by an n-tuple $(A_0, \ldots, A_n)$ of morphisms (e.g., matrices in the case of $SSE(\Lambda^+)$ and $SSE(\Lambda)$) together with elementary strong shift equivalences $(R_{ij}, S_{ji}) : A_i \to A_j$ whenever $i < j$ which satisfy the *Triangle Identities*

$$\begin{aligned} R_{ik} &= R_{ij}R_{jk} \\ S_{ji} &= R_{jk}S_{ki} \\ S_{kj} &= S_{ki}R_{ij} \end{aligned} \tag{9.5}$$

for $i < j < k$. Let $R_0 = S_{n,0}$ and $R_i = R_{i-1,i}$ for $i = 1, \ldots, n$. The Triangle Identities show that the simplex is completely determined by the (n+1)-tuple $(R_0, \ldots, R_n)$. The cyclic operator t_n on n-simplices is given by

$$t_n(R_0, \ldots, R_n) = (R_n, R_0, \ldots, R_{n-1}) \tag{9.6}$$

It is interesting that the SSE spaces, which appeared in a totally different context, do turn out to have cyclic structures.

10. TQFT and SSE

P. Gilmer has shown in [G2] how the idea of strong shift equivalence over Z^+ and Q^+ arises in topological quantum field theory invariants (e.g., for knots). Related work has been done by D.Silver and S.Williams [SW1,SW2] more in the context of combinatorial group theory.

We begin this section by recalling the simplest axioms for a *topological quantum field theory.* See [At] and [Q] for a number of examples. Let Λ be a ring. A d-dimensional TQFT assigns to each smooth manifold M^d a Λ-module $V(M)$ and to each smooth cobordism X^{d+1} from M^d to N^d (i.e., $\partial X^{d+1} = M^d \cup N^d$) a Λ-module homomorphism $Z_X : V(M) \to V(N)$ satisfying the following properties:

(1) *Composition* Suppose X^{d+1} is a cobordism from M^d to N^d and Y^{d+1} is a cobordism from N^d to P^d. Form the union $W^{d+1} = X^{d+1} \cup Y^{d+1}$ along the common part N^n. This is a cobordism from M^d to P^d, and

$$Z_W = Z_X Z_Y$$

where composition of left Λ-module homomorphisms is read from left to right.

(2) *Naturality* A diffeomorphism $F : M^d \to N^d$ induces an isomorphism $F_* : V(M) \to V(N)$ of Λ-modules. Moreover, if $G : N^d \to P^d$ is another diffeomorphism, then

$$(GF)_* = F_* G_*$$

Also , if $F : (X^{d+1}; M^d, N^d) \to (Y^{d+1}; P^d, Q^d)$ is a diffeomorphism of cobordisms, then

$$Z_X = F_* Z_Y F_*^{-1}$$

(3) *Identity* $Z(M \times [0,1])$ = Identity from $V(M)$ to $V(M)$

(4) *Nontriviality* Let ϕ be the empty set and also the empty cobordism. Then

$$V(\phi) = \Lambda \text{ and } Z_\phi = 1$$

The manifolds and cobordisms often have additional structure such as orientability, in which case there usually involutory and Hermitian axioms as well.

A very simple example is the *Euler characteristic* TQFT. Let $\Lambda = Z[t, t^{-1}]$ and then let

$$V(M) = \Lambda \text{ for each } M$$

$$Z(X) = \text{multiplication by } t^{e(X,M)}$$

where $e(X, M) = \sum_i \ (-1)^i rank H_i(X, M)$.

Gilmer's work uses the *total finite homotopy* TQFT discussed in [Q]. The setting in [G2] is to start with a pair (M, χ) where M is a connected manifold of dimension $d+1$ and χ is a primitive cohomology class in $H^1(M)$, i.e., the homomorphism $\chi : H_1(M) \to Z$ is onto. For simplicity, assume ∂M is empty. Select a Seifert surface Σ for (M, χ); namely, $\Sigma = \theta^{-1}(p)$ where $\theta : M \to S^1$ is a smooth function representing χ and p is not a critical value of θ. Let M_∞ denote the infinite cyclic cover of M with respect to the primitive class χ, and let T be the generator of the infinite cyclic group of covering transformations acting on M_∞. Lift the Seifert surface Σ up to M_∞ and consider the cobordism E in M_∞ from Σ to $T(\Sigma)$ as in Figure 1 below. As in [Q] we let

$$V(\Sigma) = Q[\Sigma, BG]$$

be the vector space over Q with (finite) basis consisting of the set $[\Sigma, BG]$ of homotopy classes of continuous maps from Σ to the classifying space BG of the finite group G. Select an ordering of the finite set $[\Sigma, BG]$ which has, say, cardinality n. The cobordism E from Σ to $T(\Sigma)$ produces a homomorphism

$$Z_E : V(\Sigma) \to V(T(\Sigma))$$

defined by

$$Z_E([f]) = \sum_{[F]\epsilon[E,f;BG]} \#^\pi(Map_f(E, BG), F)[F|T(\Sigma)] \qquad (10.1)$$

where a representative $f : \Sigma \to BG$ of each homotopy class $[f]$ of mappings from Σ to BG is chosen, where the sum is over the finite set $[E, f; BG]$ of homotopy classes of maps $F : E \to BG$ with $F|\Sigma = f$, and where the *homotopy order* is

$$\#^\pi(Map_f(E, BG), F) = (\#\pi_1)^{-1}(\#\pi_2)(\#\pi_3)^{-1} \cdots$$

$Map_f(E, BG)$ is the space of continuous mappings $F : E \to BG$ with $F|\Sigma = f$. The homotopy groups are those of the component of $Map_f(E, BG)$ containing F. Each π_i is a finite set, and $\#\pi_i$ denotes the number of elements in π_i. The product is finite because $Map_f(E, BG)$ has only finitely many nontrivial homotopy groups. The homeomorphism T gives an isomorphism from $V(T(\Sigma))$ back to $V(\Sigma)$ by means of which Z_E becomes an endomorphism $A = A(\Sigma)$ of $V(\Sigma)$. With respect to the basis for $V(\Sigma)$ chosen above, we see that

(10.2) A is an $n \times n$ matrix with coefficients in Q^+

The rational Turaev-Viro module [G1] is nothing other than the rational dimension group $(G_A \otimes Q, s_A)$.

Since M is connected, it is possible to choose a connected Seifert surface Σ. In this case it follows from (1) of Exercise 4.13 in [Q] that

(10.3) $A(\Sigma)$ is a matrix with entries in Z^+

If $d = 0$, then $M = S^1$, Σ is a point, and $A = \{1\}$. In [G2] Gilmer proves

THEOREM 10.4. *If $d \geq 2$, then the strong shift equivalence class of $A(\Sigma)$ over Z^+ is independent of the choice of the connected Seifert surface Σ and the choices made in constructing $A(\Sigma)$.*

The theorem is also true for $d = 1$, although Gilmer's proof in this case is different and relies on the elementary topological classification of Riemann surfaces.

Here is Gilmer's idea for showing the strong shift equivalence class of $A(\Sigma)$ is well defined over Z^+ when $d \geq 2$. Let Σ and Σ' be disjoint Seifert surfaces arranged inside M_∞ as in Figure 2 below. The matrix $A = A(\Sigma)$ comes from the cobordism from Σ to $T(\Sigma)$. The matrix $A' = A(\Sigma')$ comes from the cobordism from Σ' to $T(\Sigma')$. Let R be the matrix obtained by using the formula (10.1) with $T(\Sigma)$ replaced by Σ' and E replaced by the cobordism X from Σ to Σ' . Let S be the matrix obtained by using the formula (10.1) with Σ replaced by Σ' and E replaced by the cobordism Y from Σ' to $T(\Sigma)$. It follows from the TQFT composition property applied to the unions $X \cup Y$ and $Y \cup T(X)$ that

(10.5) $$A = RS \qquad \text{and} \qquad SR = A'$$

If $d \geq 1$, Gilmer observes that any two Seifert surfaces for χ are related by isotopies combined with moves between pairs of disjoint Seifert surfaces as in Figure 2. So the strong shift equivalence class of A over Q^+ is well defined. If $d \geq 2$, this can be done in such a way that all the intermediate Seifert surfaces and cobordisms are connected. The argument is like the proof of stable equivalence of Seifert surfaces well known to knot theorists. See [Li, p.79]. Hence, it follows from (1) of Exercise 4.13 in [Q] that all the strong shift equivalences in (10.5) are over Z^+ when $d \geq 2$.

A guiding philosophy which led to the development of the SSE spaces is that if objects in a given situation are related to K_1, then automorphisms should be related to K_2. In the present context this says the next step is to study the connection between diffeomophisms and symmetries of shifts. For example, let $Diff_\chi(M,p)$ denote the group of diffeomorphisms of M which fix a basepoint p and preserve χ. Work in progress shows there is a homomorphism

(10.6) $$\pi_0 Diff_\chi(M,p) \to \pi_1(SSE(Q^+), A)$$

There are other variations depending on the choice of TQFT and relaxing the condition that p be fixed. This is accomplished by using the SSE spaces together with one and two parameter families of functions from M to S^1 in a similar fashion to pseudo-isotopy theory [C,HW]. The method also uses the space of Seifert surfaces, which are analogous to cross sections of flows and Markov partitions.

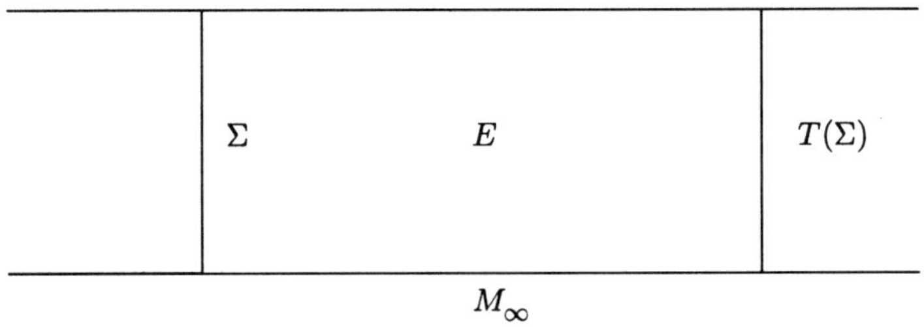

Figure 1

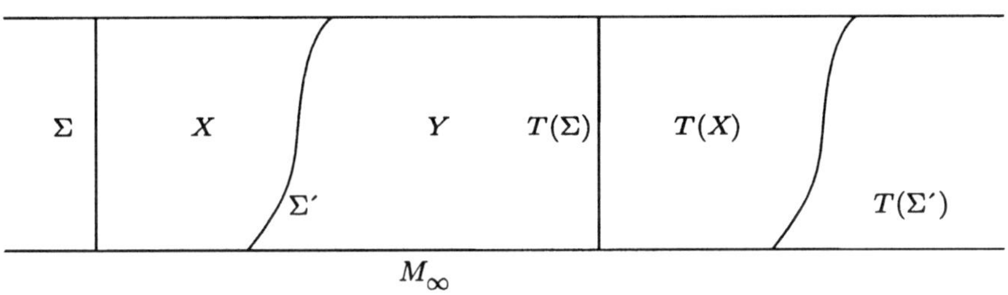

Figure 2

References

[A] R.L. Adler, *The torus and the disk*, IBM Journal of Research and Development, Vol.31, No.2, March 1987,224-234

[At] M. Atiyah, *Topological quantum field theories*, Pub. Math. IHES No.68,1988, 175-186

[AKM] R.L. Adler, A. Konheim, and M. McAndrew, *Topological entropy*, TAMS,114, 1965,309-319

[B] K. Baker, *Strong shift equivalence of* 2×2 *matrices of non-negative integers*, ETDS,1983, 3, 501-508

[Ba] L. Badoian, *Flow equivalence of subshifts of finite type and K-theory*, PhD thesis, 1998, UC Berkeley

[Bl] S. Bloch, *Algebraic K-theory and crystalline cohomology*, Pub. Math. IHES, 47, 1977, 197-238

[Bo1] M. Boyle, *The stochastic shift equivalence conjecture is false*, Contemporary Mathematics, Vol.135, 1992, 107-110

[Bo2] ________, *Symbolic dynamics and matrices in combinatorial and graph-theoretic problems in linear algebra*, IMA Volumes in Mthematics and Its Applications,Vol.50 (1993), editors R.A. Brualdi, S. Friendlander, and V. Klee, 1-38

[Bo3] ________, *Positive K-theory in symbolic dynamics*, to appear in Dynamics and Randomness, editors A. Maass, S. Martinez, and J. San Martin, Kluwer, 2002, 31-52

[Bo4] ________, *Flow equivalence of shifts of fiite type via factorizations*, to appear in Pacific Journal of Mathematics

[Bow] R. Bowen, *Markov partitions for Axiom A diffeomorphisms*, Amer. Jour. Math. 92, 1970, 725-747

[BF] M. Boyle and U.Fiebig, *The action of inert finite order automorphisms on finite subsystems*, ETDS 11, 1991, 413-425

[BGMY] L. Block, J. Guckenheimer, M. Misiurewicz, L.-S. Young, *Periodic points and entropy of one dimensional maps*, Springer-Verlag LNM No. 819,18-34

[BH1] M. Boyle and D. Handelman, *Algebraic shift equivalence and primitive matrices*, Trans. AMS 336, No. 1, 1993, 121-149.

[BH2] ____________, *The spectra of nonnegative matrices via symbolic dynamics*, Annals of Mathematics, 133, 1991, 249-316

[BK1] M. Boyle and W. Krieger, *Periodic points and automorphisms of the shift*, Trans. AMS 302 , 1987, 125-149

[BK2] ____________, *Automorphisms and subsystems of the shift*, Jour. Reine Angew. Math. 437, 1993, 13-28

[BL] M. Boyle and D. Lind, *Small polynomial matrix representations of nonnegative matrices*, to appear in Linear Algebra and Its Applications

[BLR] M. Boyle, D. Lind, and D. Rudolph, *The automorphism group of a shift of finite type*, Trans. AMS 306, 1988, 71-114

[BMT] M. Boyle, B. Marcus, and P. Trow, *Resolving maps and the dimension group for shifts of finite type*, Mem. Amer. Math. Soc. vol.70 (1987), Number 377, Providence, R.I.

[BW] M. Boyle and J.B. Wagoner, *Positive algebraic K-theory and shifts of finite type*, preliminary manuscript , UC Berkeley, 2001

[BaW] L. Badoian and J.B. Wagoner, *Simple connectivity of the Markov partition space*, Pac. Jour. Math. **193**,2000, No.1, 1-3

[C] J. Cerf, *La stratification naturelle des espaces de fonctions différentiables réelles et le theoreme de la pseudo-isotopie*, Publ. Math. IHES, No. 39, 1970, 5-173

[CK1] J. Cuntz and W. Krieger, *Topological Markov chains and dicyclic dimension groups*, J. Reine Angew. Math. 320, 1980, 44-51

[CK2] ____________, *A class of C*-Algebras and topological Markov chains*, Inventiones Math. 56, 1980, 251-268

[E] E.G. Effros, *Dimensions and* C^* *Algebras*, CBMS No. 46, AMS,1981

[F] U. Fiebig, *Gyration numbers for involutions of subshifts of finite type I*, Forum Math.4, 1992, 77-108 and II, Forum Math. 4, 1992, 183-211

[Fr] J. Franks, *Homology and dynamical systems*, CBMS No. 49, AMS,1982

[G1] P.M. Gilmer, *Invariants for one-dimensional cohomology classes arising from TQFT*, Topology and its Applications 75, 1997,217-259

[G2] ________, *Topological quantum field theory and strong shift equivalence*, preprint from Louisiana State University,1997, to appear in Bulletin of the Canadian Mathematical Society

[GR] S. Geller and L. Roberts, *K_2 of some truncated polynomial rings* , Ring Theory Waterloo, Springer-Verlag LNM 734, 249-278

[H] G. Hedlund, *Endomorphisms and automorphisms of shift dynamical systems*, Math. Systems Theory **3**, 1969, 320-375

[HW] A. Hatcher and J.B. Wagoner, *Pseudo Isotopies of Compact Manifolds*, Astérisque No. 6, Société Mathématique de France, 1973

[K] B. Kitchens,*Symbolic dynamics. One-sided,two-sided and countable state Markov shifts*, Springer-Verlag, 1998

[Ke] F. Keune, *The relativization of K_2*, Journal of Algebra,1978, 159-177

[Kol] A.N. Kolomogorov, *New metric invariants of transitive dynamical systems and automorphisms of Lebesgue spaces*, Dokl. Akad. Nauk. SSSR,119, 1958, 861-864

[Kr] W. Krieger, *On dimension functions and topological Markov chains*, Invent. Math. 56, 1980, 239-250

[KOR] K.H.Kim, N. Ormes, and F.W.Roush, *The spectra of nonnegative integer matrices via formal power series*, Jour. Amer. Math. Soc. **13** ,2000, No.4, 773-806

[KR1] K.H.Kim and F.W.Roush, *Some results on decidability of shift equivalence*, J. Comb. Inform. System Sci. 4, 1979, 123-146.

[KR2] ________________, *Decidability of shift equivalence*, in J.C. Alexander, ed., Dynamical Systems, Lecture Notes in Mathematics, vol. 1342, Springer-Verlag, Heidelberg, 1988.

[KR3] ________________, *On the structure of inert automorphisms of subshifts*, Pure Mathematics and Applications 2, 1991, 3-22.

[KR4] ________________, *Williams's conjecture is false for reducible subshifts*, Jour. AMS 5, 1992, 213-215.

[KR5] ________________,*Williams's conjecture is false for irreducible subshifts*, ERA AMS 3, 1997, 105-109

[KR6] ________________, *Williams's conjecture is false for irreducible subshifts*, preprint from Alabama State University, 1997, to appear in Annals of Mathematics

[KR7] ________________, *Path components of matrices and strong shift equivalence over Q^+*, Linear Algebra and Applications, 145,1991,177-186

[KR8] ________________, *Strong shift equivalence of Boolean and positive rational matrices* , Linear Algebra and Applications, 161,1992,153-164

[KRW1] K.H. Kim, F.W. Roush, and J. Wagoner, *Automorphisms of the dimension group and gyration numbers*, Jour. AMS 5, 1992,191-212.

[KRW2] ________________________,*Characterization of inert actions on periodic points, Part I and Part II*,Forum Mathematicum **12** ,2000, 565-602 and 671-712

[KRW3] ________________________ , *Inert Actions on Periodic Points*, ERA AMS 3, 1997, 55-62

[KRW4] __________, *The Shift Equivalence Problem*, The Mathematical Intelligencer, Vol. 21, No. 4, 1999, 18-29

[Li] W.B.R. Lickorish *An Introduction to Knot Theory*, GTM 175, Springer-Verlag, 1997

[Lo] J.- L. Loday, *Cyclic Homology*, Springer-Verlag, 1998

[LM] D. Lind and B. Marcus,*An introduction to symbolic dynamics and coding*, Cambridge University Press,1995

[M] B. Marcus, *Symbolic dynamics and connections to coding theory, automata theory and system theory*, AMS Proc. Symp. Applied Math., 50,1995, 95-108.

[Mi] J. Milnor, *Introduction to Algebraic K-Theory*, Annals of Mathematics Studies Number 72, Princeton University Press, 1971

[MB] S. MacLane and G. Birkhoff, *Algebra*, MacMillan Company, Second Edition, 1979

[MH] M. Morse and G.A. Hedlund, *Symbolic Dynamics*, Amer. Jour. Math. 60(1938),815-866.

[MS] H. Maazen and J. Stienstra, *A presentation for K_2 of split radical pairs*, Journal of Pure and Applied Algebra 10, 1977, 271-294

[MT1] B. Marcus and S. Tuncel, *Entropy at a weight-per-symbol and an imbedding theorem for Markov chains*, Inventiones Mathematicae, 102, 1990, 235-266.

[MT2] ____________________, *The weight-per-symbol polytope and scaffolds of invariants associated with Markov chains*, Ergodic Theory and Dynamical Systems, 11, 1991, 129-180.

[MT3] ________, *Matrices of polynomials, positivity, and finite equivalence of Markov chains*, AMS Journal, 6, 1993, 131-147

[N] M. Nasu, *Topological conjugacy for sofic systems and extensions of automorphisms of finite subsystems of topological Markov shifts*, Springer Lecture Notes in Math 1342, 1988

[PT] W. Parry and S.Tuncel, *Classification problems in ergodic theory*, LMS Lecture Notes 67, Cambridge University Press, 1982

[PW] W. Parry and R.F. Williams, *Block coding and a zeta function for Markov chains*, Proc. LMS 35,1977, 483-495

[Q] F. Quinn, *Lectures on axiomatic topological quantum field theory*, Geometry and Quantum Field Theory (edited by D. Freed and K. Uhlenbeck), IAS/Park City Mathematics Series, Vol. 1, 1995, 325-453

[R] J. Rosenberg, *Algebraic K-Theory and Its Applications*, GTM 147, Springer-Verlag, 1994

[S] S. Smale, *Differential Dynamical Systems*, BAMS, Vol. 73, No. 6, 1967, 747-847

[Sh] C. Shannon, *A mathematical theory of communication*, Bell Sys.Tech Jour., 27,1948, 379-423, reprinted by University of Illinois Press,1963

[St] J. Stienstra, *On K_2 and K_3 of truncated polynomial rings*, LNM 854, Springer,1981

[SW1] D. Silver and S. Williams, *Augmented group systems and subshifts of finite type*, Israel Journal of Mathematics 95, 1996, 231-251

[SW2] ________, *Knot invariants from symbolic dynamical systems*, Trans. AMS 351,No. 8, 1999, 3243-3265

[T] S. Tuncel, *Faces of Markov chains and matrices of polynomials, Contemporary Mathematics*, 135, 1992, 391-422

[vdK] W. van der Kallen, *Le K_2 des nombres duaux*, C.R. Acad. Sc. Paris, Série A,t. 273 , 20 décembre 1971,1204-1207

[W1] J.B. Wagoner, *Markov partitions and K_2*, Publ. Math. IHES, No. 65, 1987, 91-129

[W2] ________, *Triangle identities and symmetries of a subshift of finite type*, Pacific Jour. Math. 144, 1990, 181-205

[W3] ________, *Higher dimensional shift equivalence is the same as strong shift equivalence over the integers*, Proc. A.M.S. 109, 1990, 527-536

[W4] ________, *Eventual finite order generation for the kernel of the dimension group representation*, Trans. A.M.S. 317, 1990, 331-350

[W5] ________, *Classification of subshifts of finite type revisited*, Contemporary Mathematics, Vol.135, 1992, 423-436

[W6] ________, *Markov chains,C^*-algebras, and K_2*, Advances of Mathematics, Vol.71, No.2,1988,133-185

[W7] ________, *Strong shift equivalence theory and the shift equivalence problem*, Bull. Amer. Math. Soc., Vol. 36, No. 3, 1999, 271-296

[W8] ________, *Strong shift equivalence and K_2 of the dual numbers*, Jour. für Reine und Angew. Math. **521**, 2000, 119-160

[Wi1] R.F. Williams, *Classification of subshifts of finite type*, Annals of Math.(2) 98,1973, 120-153; Errata ibid. 99 ,1974, 380-381.

[Wi2] ________, *Shift equivalence of matrices in GL(2,Z)*, Cont. Math. 135, AMS , 1992, 445-451

MATHEMATICS, UNIVERSITY OF CALIFORNIA, BERKELEY, CA, 94720
E-mail address: wagoner@math.berkeley.edu

Index

Titles in This Series

ISBN 0-8218-3157-7